Carolina de Oliveira Fialho Pereira
Rute T. S. Ribeiro

PATHOGENICITY AND BIOLOGICAL CONTROL OF Fusarium WITH Trichoderma

Carolina de Oliveira Fialho Pereira
Rute T. S. Ribeiro

PATHOGENICITY AND BIOLOGICAL CONTROL OF Fusarium WITH Trichoderma

Pathogenicity, antagonism, in vitro tests and vegetative compatibility groups

ScienciaScripts

Imprint

Any brand names and product names mentioned in this book are subject to trademark, brand or patent protection and are trademarks or registered trademarks of their respective holders. The use of brand names, product names, common names, trade names, product descriptions etc. even without a particular marking in this work is in no way to be construed to mean that such names may be regarded as unrestricted in respect of trademark and brand protection legislation and could thus be used by anyone.

Cover image: www.ingimage.com

This book is a translation from the original published under ISBN 978-613-9-65467-3.

Publisher:
Sciencia Scripts
is a trademark of
Dodo Books Indian Ocean Ltd. and OmniScriptum S.R.L publishing group

120 High Road, East Finchley, London, N2 9ED, United Kingdom
Str. Armeneasca 28/1, office 1, Chisinau MD-2012, Republic of Moldova, Europe
Printed at: see last page
ISBN: 978-620-7-77775-4

CONTENTS

ACKNOWLEDGMENTS

- I would like to thank my advisor, Professor Dr. Rute T. S. Ribeiro, for her patience, dedication and guidance.

- To CAPES for the grant that made this work possible.

- To my parents and siblings, who were always there for me at all times, supporting my decisions unconditionally.

- To my grandmother, for being an example of strength and perseverance.

- To my colleagues at the Biological Control of Plant Diseases Laboratory Camila, Luis, Franciele, Rosemeri, Araceli, Fabiane, Elisa, Jaqueline, and especially the laboratory technician and friend Màrcia, for all her help and free psychological consultations over a cup of coffee.

- To all the colleagues and professors I've worked with over the years, especially my colleague Carina Cassini, for her help in times of crisis and difficulty.

- Professors Mirian Salvador and Neiva Monteiro de Barros, for their corrections and suggestions, which were very important for the development of this work.

- To my friends Taisa, Marina, Annia and Bàrbara, for listening to me in times of difficulty, and for the friendship that transcends geographical barriers and time.

- To Sergio, for his patience, love and help at all times.

SUMMARY

Phytopathogenic fungi belonging to the genus *Fusarium* are known to cause plant diseases in various hosts. Among these is the tomato plant, which is attacked by the three known races of *Fusarium oxysporum* f. sp. *lycopersici* (*Fol*), which cause vascular wilt. To control this disease, the use of microorganisms, such as antagonistic isolates of *Trichoderma* sp. can be an alternative to the use of agrochemicals. In the present study, the pathogenicity of ten isolates of *Fol* was evaluated on differentiated tomato cultivars, with only four proving to be pathogenic to susceptible cultivars, and none of them was diagnosed as belonging to race 3. The results obtained in the direct confrontation tests with the *Trichoderma* sp. isolates showed high variability in terms of mycoparasitic capacity, with the best overlapping results for isolates T3, T8 and T17. In these cases, evidence of host colony overlap was observed for at least six of the nine antagonist isolates. In relation to the inhibitory capacity of these isolates, the results of the greatest inhibition of the growth of the phytopathogen colony were obtained for isolates T2 and T3. *Fol* strain 34970 was the most resistant to the antagonistic actions of the *Trichoderma* sp. isolates, and isolate TO 11 had the highest inhibition averages. In the test for the production of volatile metabolites, only the isolates Fusarium 23, Fusarium 27 and 34970 de *Fol* were inhibited after 120 hours of testing. In the biological control of tomato fusariosis, caused by *Fol isolate* TO 245, the seedlings treated with *Trichoderma* sp. isolates T6 and T17 showed a lower incidence of disease, with scores of 1.25 and 0.75 respectively compared to 1.87 in the control group, according to the disease severity scale of Vakalounakis *et al.* (2004). Treatments with isolates T6 and T17 on substrate contaminated with *Fol* isolate 1205/2 showed a value of 0.50 and 0.38, according to the scale, compared to 0.50 for the control treatment. The average heights, lengths and dry weights of the roots showed no statistical difference for any of the treatments, but the dry weight of the aerial part was significantly higher for the plants treated with the T6 isolate in substrate infested with the 1205/2 *Fol* isolate. All the pathogenic isolates paired with the standard strains of the vegetative compatibility groups showed signs of heterokaryosis with the standard strain 34970 corresponding to GCV 0030, indicating a possible genetic similarity between the isolates used.

KEYWORDS: *Fusarium oxysporum* f. sp. *lycopersici*; biological control; tomato; vegetative compatibility groups.

1. INTRODUCTION

Cosmopolitan fungi of the genus *Fusarium* Snyder & Hansen are known to cause diseases in a wide variety of plants. Fusarium is a vascular disease that causes an interruption in the translocation of ascending sap, a reduction in shoot growth, wilting of leaves and fruit and internal blackening of the stem. Infected plants can die suddenly and usually in clusters. The damage caused by fusarium wilt is due to the infection and death of seedlings, the death of plants during vegetative development and the rotting of fruit, which loses its commercial value. The fungus attacks the plant from the ground, penetrates its roots and invades its vascular system, causing the older leaves to bend downwards.

Among the host plants of *Fusarium* spp. is the tomato plant, whose cultivation occupied an area of more than 3.7 million hectares worldwide in 2001, with a production of approximately 100 million tons. Brazil is one of the top ten producing countries, having produced a total of 3,641,400 tons of tomatoes in 2003. Fusarium wilt is one of the most widespread diseases in the countries where this vegetable is grown. The use of implements for soil management and weed control injures the roots, causing openings that favor infection by the fungus. Poorly drained soils and especially soils with an excess of organic matter favor the occurrence of the disease. High temperatures and low relative humidity increase its severity.

Physical, cultural, chemical and biological control methods are used to stop the pathogen. One of the most widely used control methods until recently was soil fumigation with methyl bromide. This method is efficient for disinfestation, but it is one of the methods responsible for depleting the ozone layer, which is why restrictions are imposed on its use. In Brazil, the commercialization of methyl bromide for soil disinfestation is prohibited. Currently, the technique used to control fusarium is based on the use of cultivars genetically modified for resistance to the different races of *Fusarium oxysporum* f. sp. *lycopersici* (Sacc.) W.C. Snyder & H. N. Hans.

Among the alternatives to conventional control methods is solarization, which consists of covering the wet soil with transparent polyethylene during the months when the sun shines the most, with the aim of increasing its temperature and destroying fungal propagules. However, this method can be expensive and most producers prefer to plant another crop in the soil. Another technique is microbiological control of the pathogen with antagonistic microorganisms to keep the pathogen population below levels that can cause economic damage to the crop.

Fungi and bacteria are the most widely used microorganisms for biological control. Among fungi, some species of the genus *Trichoderma* sp. Person have stood out as antagonists of a range of pathogens that cause diseases in various crops, including *Fusarium* sp. These biocontrol species have a broad spectrum of action, acting in various ways at the plant/phytopathogen interface and indirectly controlling various diseases.

Among the various forms of action, the antagonistic species of *Trichoderma* sp. develop mycoparasitism, antibiosis, competition for nutrients, induction of resistance and solubilization of nutrients for plants. Another relevant characteristic is that they are highly resistant to a wide range of toxic substances, which leads to the hypothesis that they can be used in Integrated Pest Management (IPM) systems.

In order to control diseases efficiently, knowledge of the causal agent is necessary. Molecular and vegetative compatibility techniques can be used to characterize isolates and carry out genetic similarity analyses, as well as helping to discover ways of introducing diseases into certain regions.

Bearing in mind that in Rio Grande do Sul there is a high incidence of tomato fusariosis, we propose characterizing the pathogenicity of some isolates, carrying out direct confrontation tests and evaluating the effect of volatile metabolites from *Trichoderma* sp. isolates against isolates of Fusarium f. sp. lycopersici, as well as characterizing the vegetative compatibility groups of isolates of Fusarium f. sp. lycopersici. against isolates of *Fusarium* f. sp. *lycopersici,* and characterization of the vegetative compatibility groups of isolates of *Fusarium* f. sp. *lycopersici, as* well as microbiological control of tomato wilt using antagonistic isolates of *Trichoderma* sp.

2. LITERATURE REVIEW

2.1 The tomato plant (*Lycopersicon esculentum* Mill.) and the fungus *Fusarium oxysporum*

Tomatoes are the third largest crop in Brazil. Around 3 million tons are sold annually, 80% of which are in the states of Sao Paulo, Minas Gerais, Rio Grande do Sul, Rio de Janeiro and Goiás. Table tomato cultivation is highly sensitive to pests and diseases, requiring intensive use of chemical pesticides, which pose great risks of contamination to workers, consumers and the environment in general (Santos & Noronha, 2001).

Global tomato production has doubled in the last 20 years, with one of the main factors behind the crop's expansion being the growth in consumption, reinforced by the demand for healthier foods. Between 1985 and 2005, world *per capita* tomato production grew by around 36%, from 14 kg to 19 kg per person per year (Simao & Rodriguez, 2008).

According to IBGE data, the average yield of this crop is around 57,435 kg/ha of cultivated area, a productivity that has been increasing over time due to advances in phytosanitary techniques (Table 1). Exports of fresh tomatoes, prepared or preserved tomatoes and tomato sauces in 2005 amounted to 476,9809 and 2,585 tons, respectively. The main importing countries were Argentina, Paraguay and Uruguay.

Fusarium oxysporum Snyder & Hansen, is a fungus commonly found in soil, where it survives in the form of chlamydospores and grows saprophytically in organic matter. This species also includes many plant pathogens that can induce necrosis or wilting in crops of economic importance.

Table 1: Tomato production, area and productivity in Brazil over the years (IBGE, 2005).

Year	Production (kg)	Area harvested (1000. ha)	Average yield (kg/ha)
1990	2.261	61	37.143
1991	2.344	61	38.510
1992	2.141	52	41.014
1993	2.348	54	43.706
1994	2.689	62	43.407
1995	2.715	62	43.752
1996	2.649	71	37.317
1997	2.718	65	41.781
1998	2.784	64	43.569
1999	3.305	66	50.356
2000	3.005	57	52.976
2001	3.103	57	53.980
2002	3.653	63	58.428
2003	3.709	63	58.423
2004	3.516	60	58.445
2005	3.155	55	57.435

Tomato fusarium wilt is caused by the fungus *F. oxysporum* f. sp. *lycopersici* (Fol) and occurs in all regions of Brazil. In the 1970s, in areas where cultivation had already been established for a longer period of time, it was common for 100% of the cultivated plants to be destroyed or for the harvest period to be drastically reduced due to premature plant death (Galli, 1980).

Tomato plants with symptoms of fusarium wilt their upper leaves, especially during the hottest hours of the day. The older leaves turn yellow and often wilting or yellowing is observed on only one side of the plant or leaf (Figure 1). The fruits do not develop, they ripen when they are still small and production is reduced. When the stem is cut close to the roots, necrosis of the vascular system (xylem) can be seen. High temperatures (around 28 °C), sandy soils with low pH and nematode attacks favor the disease. The fungus survives in the soil for more *than* seven years, mainly by means of microsclerotia (the fungus's resistance structure) (Embrapa, 2008).

Figure 1: Tomato plant showing symptoms of fusarium wilt.

In Brazil, fusarium wilt was once an important disease for staked tomato plants, when susceptible varieties were planted. With the advent of cultivars resistant to races 1 and *2,* the disease became less important in the late 1970s. In recent years, however, with the emergence of a third race of this pathogen, efficient and safe control measures have become a priority in tomato cultivation (Kimati *et al.* 2005).

In certain regions of China (Zhejiang), fusarium wilt has become a serious problem, causing the death of 30 to 60% of tomato plants grown in greenhouses (Zhihao *et al.* 2000). In hydroponic cultivation, wilt caused by *F. oxysporum* f. sp. *lycopersici* is one of the most damaging diseases, usually occurring in the middle to late stages of the growing season. It is difficult to control because contaminated water circulates, making it easier for the pathogen to spread to all the seedlings (Song *et al.* 2004).

In the plant disease diagnosis clinic at the Biotechnology Institute of the University of Caxias do

Sul, more than 50% of the samples of diseased plants have been diagnosed with a disease caused by *Fusarium* spp. Among the plant species that are received for analysis, the tomato plant is the second most affected, with only a smaller number of samples than the grapevine, which is a very common crop in the northeastern region of the state of Rio Grande do Sul.

2.2 Ways of controlling the disease

Among the fungicides tested by Kopacki & Wagner (2006) to control the chrysanthemum pathogen *F. avenaceum* Fr. sacc., only difenoconozole, flusilazole and carbendazim were effective at the recommended dose. Those containing captan, mancozeb and chlorotanil did not inhibit the mycelial growth of the fungi, nor did they have a lasting effect on the crop in the long term. Another aspect to be considered is that when working with chemical products, more care should be taken with regard to genetic variability in populations, since pathogenic isolates can differ in their sensitivity to fungicides.

For the control of *F. subglutinans* Wollenw. & Reinking, the only fungicides that inhibited mycelial growth from 100 ppm were tebuconazole and thiabendazole. Captan and methyl thiophanate showed inhibition of less than or equal to 85%, even when a dose of 1000 ppm was applied (Fischer *et al.* 2006).

With symptoms very similar to those produced in tomatoes, the fungus *Fusarium oxysporum* f. sp. *phaseoli* Kendrick & Snyder is one of the biggest phytosanitary problems in Colombia and the Andean region. Conventional management of this disease by farmers is difficult, requiring the application of fungicides, which sometimes prove to be inefficient, increasing production costs, contributing to environmental contamination and damaging the soil biota (Avendano *et al.* 2006).

Another example of unsuccessful control of grapevine wilt, caused by *F. oxysporum* f. sp. *herbemontis, is* considered to be one of the main causes of death in the cultivars *Vitis riparia* and *V. berlandieri.* According to Andrade *et al.* (1993), the use of chemical fungicides to control the grapevine root system pathogen has been insufficient. Therefore, one way of controlling the disease could be to use plants of resistant varieties associated with biofungicides, which could increase the efficiency of control.

Another host susceptible to the pathogen is the pineapple tree, since losses can reach up to 100% of production. The most widely planted cultivars in Brazil are susceptible to the disease, posing a serious control problem (Santos *et al.* 2002).

The cultivation of bananas (*Musa* spp.) in the state of Roraima is one of the main agricultural activities of small and medium-sized producers, with a production of around 23,720 tons/year. Among the diseases, Panama disease, caused by the soil fungus *F. oxysporum* f. sp. *cubense* (Smith)

Snyder & Hansen, is considered one of the main ones, and the use of resistant cultivars is the main control method employed (Nechet *et al.* 2004).

The control of diseases caused by *Fusarium* sp. involves various measures and cultural practices, including all those of a preventative nature such as liming, irrigation, the addition of organic matter and, above all, the use of cultivars with genetic resistance (Carvalho *et al.* 2005).

The use of resistant cultivars is one of the most widely used methods for avoiding the damage caused by tomato fusariosis. Different mechanisms have been described as possibly responsible for reducing colonization by phytopathogens in resistant cultivars, including the formation of barriers that prevent the fungus from progressing, confinement of the pathogen to the primary xylem, proliferation of pathogen-free vessels, restriction of pathogen growth and induction of phytoalexin production (De Cal *et al.* 2000).

Among the classic ways of controlling fusarium wilt, crop rotation can be cited as an example. Crop-livestock integration systems have also promoted the suppression of pathogens that survive in the soil. In this case, the suppression of pathogens is credited to the management of species of *Brachiaria* sp. which, together with the addition of organic matter to the soil and the formation of straw, stimulate the development of fungi and bacteria that reduce the inoculum of pathogens (Louzada *et al.* 2004).

The chemical control of fusarium wilt has proved to be inefficient and is one of the factors inducing researchers and farmers to look for more effective alternatives, but without causing an increase in soil and water pollution levels.

An alternative method for controlling diseases caused by *Fusarium* sp. is the use of antagonistic microorganisms, such as antagonistic bacteria present in the rhizosphere of plants. In the case of *cucumber* root rot (*Cucumis sativus* L.), caused by *Fusarium solani* (Mart.) Hans, *Bacillus subtilis* Cohn and *Pseudomonas* spp. Schroeter stand out, which can inhibit up to 78% of the pathogen's mycelial growth (Melo & Valarini, 1995).

In confrontation tests between *Trichoderma harzianum* and *F. oxysporum* f. sp. *radicis - lycopersici* (Jarvis & Shoemaker) Hibar *et al.* (2005), obtained results of more than 65% inhibition of the pathogen after a period of 4 days at 25°C. After 6 days, *T. harzianum not* only inhibited mycelial growth, but also invaded and sporulated on the colonies of the pathogen, demonstrating its capacity as a mycoparasite.

Using the same technique, Paradela (2002) evaluated *in vitro* the behavior of *Fusarium* sp., *Rhizoctonia solani* J. G. Kuhn and *Sclerotinia sclerotiorum* (Lib.) de Bary. *in* relation to 11 isolates of *Trichoderma* spp. and obtained varied results of antagonism *in* relation to the strains used. The

Trichoderma spp. isolates behaved differently in terms of aggressiveness, exerting antagonism of the antibiosis, parasitism and competition types.

Fungi belonging to this genus have already been reported successfully controlling other species of *Fusarium* sp., as in the work of Luongo *et al.* (2005) in which there was a significant reduction in the sporulation of *F. culmorum* (Wm.G. Sm.) Sacc, *F. graminearum* (Schwein.) Petch, and *F. proliferatum* (Matsush.) Nirenberg ex Gerlach & Nirenberg, when grown in the presence of the antagonist *Trichoderma,* yeasts and non-pathogenic *Fusarium.* The reduction in sporulation ranged from almost 100% to less than 20%, and many of these microorganisms were selected for further testing and potential application on corn and other plants.

As an alternative that has been used recently, the use of non-pathogenic *F. oxysporum* can reduce disease severity and maintain healthy plant growth. Tomato plants treated with spore suspensions of this non-pathogenic form before being transplanted into substrate previously infested with pathogenic isolates of *F. oxysporum* f. sp. *lycopersici,* showed lower disease severity and greater average heights compared to the control which did not receive treatment with non-pathogenic *Fusarium* sp. (Silva & Bettiol, 2005).

Larkin & Fravel (1998) used different rhizosphere bacteria and fungi to control tomato vascular wilt caused by *Fol.* hamatum. The results were satisfactory for non-pathogenic *Fusarium* (up to 100% reduction in disease incidence) and *T. hamatum* (64%), and for the combination of non-pathogenic *Fusarium* and *Pseudomonas* (60%).

Despite the promising aspect of using non-pathogenic *Fusarium* sp. to control fusarium wilt, there is still a need for more research in this area as the control mechanisms are not yet fully understood. As an example of the importance of these studies, we can cite the work of Forsyth *et al.* (2006) in which some non-pathogenic isolates of the genus suppressed the wilt caused by *Fusarium,* while others increased the development of the disease, which reflects the diversity of host responses to the stimulus generated by the fungus. However, there is no detailed explanation as to why these differences in response to disease development occur.

2. 3 *Fusarium* sp. races and vegetative compatibility

Isolates of *F. oxysporum* that are pathogenic to the same host, or have the same host range, are called *formae spéciales,* and more than 70 of these forms have already been described (Kistler, 1997).

When there is a high level of specialization, *special formae* can be associated with pathogenicity on just one host species. Many *formae speciales* can be subdivided into races based on their virulence on different cultivars of a host (Armstrong & Armstrong, 1981).

The expression of resistance to the specific race is qualitative, complete or almost complete, controlled by one or a few genes (monogenic, oligogenic) with a major phenotypic effect known as major genes or genes of main effect. The expression of these genes can also be modified by the action of other major or minor genes (epistasis), by the stage of development of the plant and by the tissue attacked (Stadnik & Talamini, 2004).

In general, these major resistance genes (*R*) operate in a gene-to-gene system proposed by Flor (1971) with the avirulence genes (*avr*) of the pathogen. This hypothesis implies that incompatibility requires a dominant resistance gene in the host and a corresponding dominant avirulence gene in the pathogen. In the absence of one of these two genes, recognition of the pathogen does not occur, the plant's defenses are not initiated and the disease occurs (Mes *et al.* 1999).

Isolates of *F. oxysporum* f. sp. *lycopersici* have been grouped into 3 races according to their ability to cause disease in different tomato cultivars, possessing different disease resistance loci according to Reis *et al.* (2004). Three resistance loci have already been identified in *Lycopersicon*. Locus I was obtained from *L. pimpinellifolium* 'PI 79532' in 1940 and controls resistance to race 1. A new resistance locus was identified and characterized in accession 'PI 126915', which is a natural hybrid between *L. esculentum and L. pimpinellifolium,* in 1945. A third race of the pathogen, capable of infecting cultivars carrying both loci I and II, was recorded in Australia (Grattidge & O' Brien, 1982), and was subsequently detected in some states in the United States (Cai *et al.* 2003) and Mexico (Valenzuela-Ureta *et al.* 1996).

In Brazil, race 3 was reported in the state of Espirito Santo in 2005, when Reis *et al.* (2005) analyzed the pathogenicity of seven isolates of *F. oxysporum* from wilted tomato plants of the varieties 'Carmen' and 'Alambra', both considered resistant to races *1* and *2*. The results of applying these isolates to varieties resistant to race *3* indicated that they were the same race of the pathogen, since there was no incidence of the disease in this variety.

The second report of this race in the country was made in 2007, in the state of Rio de Janeiro by Reis & Boiteux (2007), who tested isolates obtained from plants resistant to races 1 and 2 that showed symptoms of the disease using pathogenicity tests. Seven of these isolates were characterized as belonging to race 3, reinforcing the hypothesis that the disease was transmitted via contaminated seeds, since the same pathogen appeared almost simultaneously in two geographically isolated tomato production areas.

Currently, most of the cultivars of table and processing tomatoes used in Brazil have resistance genes for races 1 and 2 of Fol. As a result, there is constant selection pressure on the pathogen, which explains why the isolates collected most recently in RJ and ES are all race 3 (Reis *et al.* 2006). New races of *Fusarium* causing wilt that overcome the resistance of commercial cultivars continue to emerge. This may explain why tomato wilt remains a serious and persistent disease, despite the intensive efforts of geneticists to develop cultivars with resistance genes (De Cal *et al.* 2000).

Recently, analytical techniques involving physiology and vegetative compatibility have been used to study the taxonomy, phylogeny and pathogenic relationships between species. In genetics, vegetative compatibility tests have been useful for characterizing the diversity between isolates, and can differentiate pathogenic and non-pathogenic populations. In addition, vegetative compatibility serves as a polymorphic marker and can also be used for "self/ non-self" recognition, as a tool for constructing genetic maps, for isogenizing strains and for identifying whether two strains are identical (Leslie, 1996).

The vegetative compatibility technique was developed by Puhalla (1985), based on the technique developed by Cove (1976), and obtained the generation of *nit* mutants of *F. oxysporum* in a medium containing chlorate. The mutants were selected by transferring the fast-growing sections of the colonies to a medium containing nitrate as the only source of nitrogen and observing that the mutants grew sparsely and without aerial mycelium. The formation of heterocultures of mutants derived from the same parental strain was indicated by the development of dense aerial mycelium in the region where the two colonies of the *nit mutants* met, i.e. the colonies underwent anastomosis. These *nit* mutants were used to force heterokaryosis and, consequently, to test isolates of *F. oxysporum* for vegetative compatibility with each other.

For a better understanding of vegetative compatibility between *nit* mutants, the nitrate reduction pathway was well characterized in the work by Correi *et al.* (1987). According to these authors, two enzymes are required to reduce nitrate to ammonium (Figure 2), and numerous genes control nitrate assimilation and its regulation is complex. Chlorate is a nitrate analog that has been very useful for studying nitrate assimilation in fungi. The reduction of chlorate to chlorite by nitrate reductase can result in chlorate toxicity for these microorganisms. Although other modes of action are possible, in general chlorate-sensitive strains can reduce nitrate to nitrite, but chlorate-resistant strains cannot. This would explain the poor growth of the resistant mutants in minimal medium containing nitrate as the only source of nitrogen.

Studying the nitrate reduction pathway in *F. moniliforme*, Klittich & Leslie (1988) concluded that the genetic control (*nit* loci) of this species is very similar to that found in *Neurospora* and *Aspergillus,* but the regulation of nitrogen metabolism as a whole may be different.

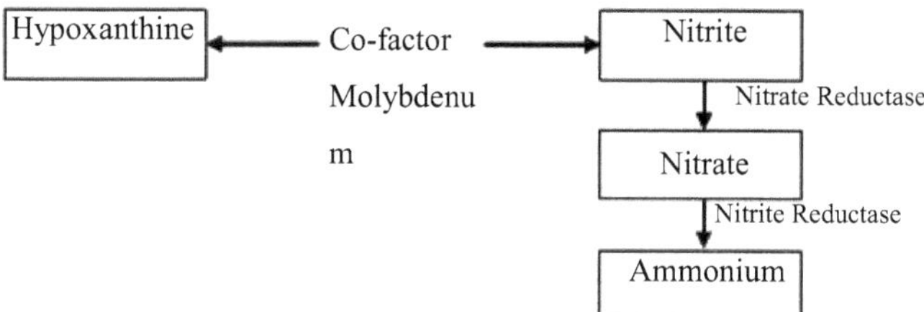

Figura 2. Nitrate and hypoxanthine utilization pathway in *Neurospora crassa* and *Aspergillus nidulans* (adapted from Correl *et al.* 1987).

When two isolates of a fungus are vegetatively compatible, their hyphae can make contact, fuse and form heterokaryons which, in many cases, are formed when identical alleles exist at each *vic/het* locus. Heterokaryosis or its opposite reaction of incompatibility have been found in many fungi including *F. oxysporum*, in which the sexual phase is not known and the exchange of genetic material occurs through parasexual mechanisms (De Oliveira & Da Costa, 2003).

The *het* genes belong to a larger group of genes responsible for the establishment and maintenance of the stable heterocyst (Figure 3). The mechanical and genetic processes that lead to *het* interaction begin before hyphae fuse, follow one or more signaling networks and culminate in the formation of a stable heterocyst or in the process of cell death (apoptosis) (Leslie & Zeller, 1996).

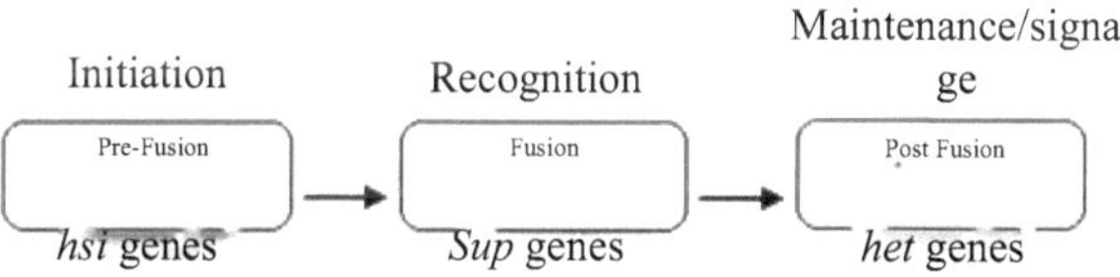

Figura 3. Simplified model identifying the steps in the vegetative compatibility interaction process and some of the genes related to this process (adapted from Leslie & Zeller, 1996).

Isolates that are or are not compatible with each other are assigned to vegetative compatibility groups (VCGs) according to their ability to form heterocompatibilities with a standard strain from each group. A high degree of homology is required for isolates to be vegetatively compatible, suggesting that pathogenic races within the same GCV may differ at a relatively small number of loci. Pathogenic diversity within a GCV indicates the prospect of potentially significant parasexual interaction, and this parasexuality may lead to the development and dissemination of new races in the future (Jacobson & Gordon, 1988).

Genetic and enzymatic analyses of *F. oxysporum* suggest that races of the same GCV have very close relationships, while isolates of the same race on different GCVs are distinct from each other. In this system, vegetative compatibility isolates are given a code consisting of four or five digits, with the first three digits corresponding to host specialization, or *formae speciales,* and the last digits corresponding to the GCVs within the *formae speciales* (Kistler *et al.* 1998).

Since the beginning of the application of this compatibility technique, *nit* mutants have been used to examine GCVs, the relationships between races and GCVs, and GCV and genetic analyses in many special forms (Ogiso *et al.* 2002). Four vegetative compatibility groups have been reported for *F. oxysporum* f. sp. *lycopersici*. Races 1 and 2 occur in GCV's 0030 to 0032 and race 3 occurs in GCV's 0030 and 0033 (Cai *et al.*, 2003).

In an attempt to determine the relationship between the GCVs and the virulence of isolates of *F. oxysporum* f. sp. *cucumerinum*, Ahn *et al.* (1998) applied the technique of Correl *et al.* (1987) to generate mutants. For this speciali form, a relationship was found between virulence (race) and the compatibility group of the isolates, proven by the anastomosis of hyphae in minimal medium (Figure 4). However, this association between race and GCV is rare in *F. oxysporum and* its cause has not yet been proven.

Among the special forms of *F. oxysporum* with GCV's already analyzed are *vasinfectum* W.C. Snyder & H.N. Hansen (Fernandez *et al.* 1994), *cepae* W.C. Snyder & H.N. Hansen (Swift *et al.* 2002), *cucumerinum* Snyder & H.N. Hansen (Vakalounakis & Fragdiaskis, 1999), *gladioli* (Massey) Snyder & H.N. Hansen (Mes *et al.* 1994), *radicis-lycopersici* (Katan *et al.* 1991), *lentis* W.L. Gordon (Belabid & Fortas, 2002) and for the species *F. graminearum* (McCallum *et al.* 2001).

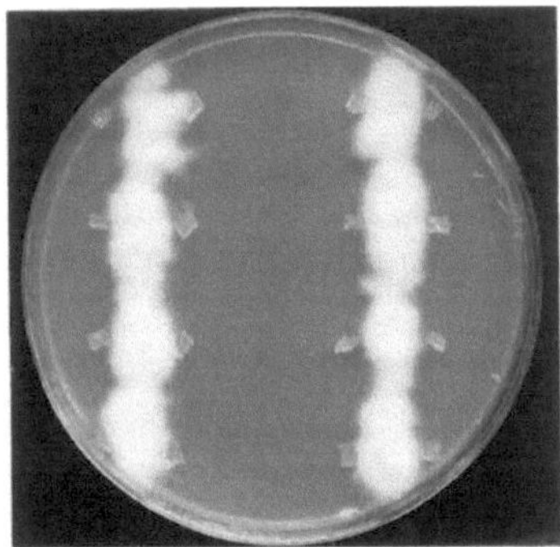

Figura 4. Anastomosis between mutants of *Fusarium oxysporum* f. sp. *cumerinum* generated in medium containing NaNO₃ as a nitrogen source (Ahn *et al.* 1998).

By 1999, 38 formae speciales had been listed in *F. oxysporum,* with 49 VCGs according to Katan & Di Primo (1999) (Table 2).

Another application of the use of GCV's is for the correct differentiation between *Fol* and *F.*

oxysporum f. sp. *radicis-lycopersici,* because in situations that are still unknown in the field, the symptoms caused by these two tomato pathogens overlap. These tests have been used and have proved to be an efficient tool for differentiating isolates (Gale *et al.* 2003).

A large number of GCVs are commonly recorded in fungi that reproduce sexually, such as *F. moniliforme* (teleomorph *Giberella fujikuroi*) J. Sheld. In contrast, fungi that reproduce exclusively asexually, such as the formae speciales of *F. oxysporum,* generally have GCVs subdivided into groups, which indicates a high degree of similarity (Vakalounakis *et al.* 2004).

Table 2. Vegetative compatibility groups of *F. oxysporum* listed according to formae specialis.

special forms	Host	Code f. sp.	Numbers assigned to the GCVs	Number of GCVs
adzukicola	*Vigna angularis*	100-	1001-1008	17
allii	*Allium sativum*	040-	10011-10019	1
asparagi	*Asparagus*	036-	0400	2
basilici	*Ocimum*	042-	0420	1
potatoes	*Ipomoea*	010-	0100, 0101, 0104	3
cepae	*Allium cepae*	012-	0120-0126; 0128-01224	24
conglutinans	*Brassica*	018-	0180-0183	6
Cuban	*Muse*	002-	0020-0022 0025, 0027, 0028	6
cucumerinum	*Cucumis sativus*	029-		2
dianthi	*Dianthus*	043-		4
erythroxyli	*Erythroxylum*	101-	1011-1013	3
fragariae	*Fragaria*	034-	0340-0345	6
garlic	*Allium sativum*	041-		1
gladioli	*Gladiolus*	019-	0190	1
lagenariae	*Lagenaria*	044-	0440	1
lilii	*Lilium*	031-		2
lini	*Linum*	003-	0030-0033	4
lupini	*Lupinus*	032-	0320	1
lycopersici	*Lycopersicon*	013-	0130-0136, 0138	8
melongenae	*Solanum melongena*	008-	0080-0082	3
melonis	*Cucumis melo*	045-	0450-0456	7
niveum	*Citrullus*	016-	0161-0168	8
opuntiarum	*Opuntia*	007-	0070-0073	5
phaseoli	*Phaseolus*	026-	0260, 0261	2
pisi	*Pisum*	009-	0090-0094 0096-0099	9
radicis-cucumerinum	*Cucumis sativus*	022-	0220	1
radicis-lycopersici	*Lycopersicon*	033-	0330-0332	3
raphani	*Raphanus*	006-	0060	1
spinaciae	*Spinacia*	023-	0230	1
tracheiphilum	*Vigna unguiculata*	020-	0201, 0202	2
tulipae	*Tulip*	011-	0110-0119 01111-01112	12
vanillae	*Vanilla*	046-		
vasinfectum	*Gossypium*			

GCV- vegetative compatibility group f. sp.- formae speciales

2.4 Biocontrol mechanisms employed by *Trichoderma* sp.

Fungi used in biocontrol, belonging to the genus *Trichoderma,* have the ability to interact parasitically and symbiotically with different substrates and living organisms, including plants and microorganisms. They can use a variety of nutrient sources and are among the most resistant to the antagonistic action of microbes, toxins and chemical substances, both natural and man-made, as well as being able to degrade some of these molecules (Woo *et al.* 2006).

As an example of this resistance, *Trichoderma* species can produce enzymes that degrade cyanide and allow normal plant growth in soils with concentrations of 50 or 100 ppm of this compound (Harman, 2006).

The mechanisms by which these fungi act are varied and can occur through the production of volatile or non-volatile metabolites, competition for nutrients and space, mycoparasitism, increased tolerance to plant stress through increased rooting and growth promotion, solubilization and sequestration of inorganic nutrients and inactivation of pathogen enzymes (Harman, 2000).

Antibiosis occurs during interactions between *Trichoderma* species and other fungi, involving diffusible low molecular weight compounds or antibiotics that inhibit the mycelial growth of target fungi.

Dennis & Webster (1971a and b) reported that *Trichoderma* species produce volatile and non-volatile compounds capable of inhibiting the mycelial growth of a variety of fungi, and that this production of antifungal substances varies according to the isolate. The production of secondary metabolites is strain-dependent and includes antifungal substances belonging to a variety of classes of chemical compounds. They have been classified according to Ghisalberti & Sivasithamparam (1991) into three categories: (i) volatile antibiotics and isocyanide derivatives; (ii) water-soluble compounds; and (iii) "peptaibols", which are linear oligopeptides of 12 to 22 amino acids.

These metabolites (Figure 5) include harzianic acid, alamethicins, tricolines, massoilactone, harzianopyridone, glisoprenins, heptelidic acid (Benitez *et al.* 2004), paracelsin, trichopoline A and B, trichorzianin, trichotoxin A, B and AB40 (Shaw & Taylor, 1986), anthraquinones, butenolides, izonotrins (Ghisalberti & Sivasithamparam, 1991), gliotoxin and gliovirin (Howell, 2003), and trichhoviridin, koninginin A, viridiol, viridin, harzianodione (Vinale *et al.* 2008). Among the most studied metabolites are harzianolide and 6-n pentyl-2H-pyran-2-one and its analog 6-n pentenyl-2H-pyran-2-one, present in several *Trichoderma* species and responsible for the coconut aroma associated with fungi of this genus, as well as showing good activity against sclerotia-forming pathogens.

According to Howell (2006), the antibiotic gliotoxin is produced by "Q" strains of *T. virens*, which have a spectrum of action against bacteria, actinomycetes and fungi, and acts synergistically with chitinases in antifungal activities. However, the antibiotic gliovirin produced by "P" strains of this same fungus has a more limited spectrum of action and is not effective against bacteria, actinomycetes and most fungi, but is a potent inhibitor of oomycetes such as *Pythium* spp. Pringsheim and *Phytophthora* spp. (Mont.) De Bary.

In the work of Küçuk & Kivanç (2003), who evaluated the effect of volatile and non-volatile metabolites of *T. harzianum* against various phytopathogenic fungi, it was found that the non-volatile metabolites were more effective at inhibiting the growth of *F. oxysporum*, *R. solanii* and *Sclerotium rolfsii* Sacc. than the volatile ones.

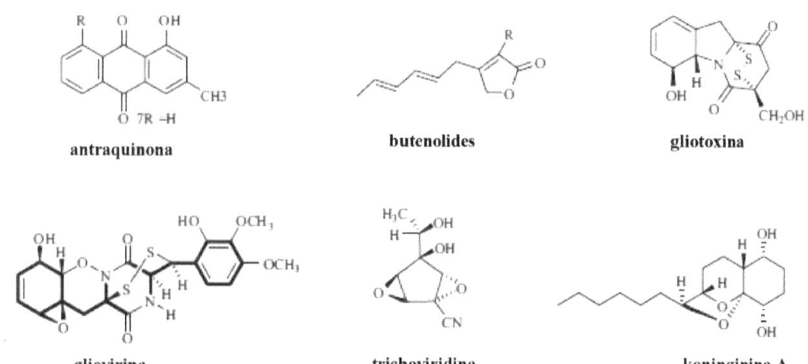

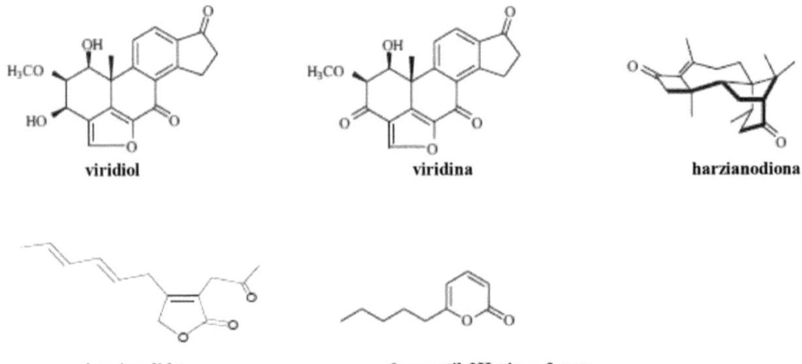

Figura 5. Chemical structure of some compounds produced by *Trichoderma* spp. (adapted from Vinale *et al.* 2005, 2008).

The same authors studied the interactions between *T. harzianum* antagonist strains and plant pathogens present in the soil (Küçük and Kivanç, 2004). The conclusion was that the metabolites produced by the antagonist inhibit the growth of phytopathogens such as *Fusarium* sp.

Antibiosis precedes mycoparasitism and colonization of the host, causing various effects on host cells, such as: retraction of the plasmalemma, breakdown of organelles, disintegration of the cytoplasm and loss of turgor (Bélanger *et al.* 1995).

The events that lead to mycoparasitism (Figure 6) take place in a series of steps that begin with the detection of other fungi and the consequent growth around them by the antagonist. After initial contact, *Trichoderma* sp. attaches itself to the host, growing spirally around the hypha and forming appendages on the surface of the target hypha (Figure 6a). At the end of this stage, it produces a series of cell wall degrading enzymes and probably some antibiotics, mainly at the interface between the appendage and the target fungal hyphae. The combined effects of these activities result in the dissolution of the cell wall and parasitism of the target fungus (Figure 6b). In the region where the appendage attaches, holes are produced with direct entry of the *Trichoderma* hyphae into the host lumen (Harman *et al.* 2004) (Figure 6c).

A partial list of pathogenic fungal genera that can be affected by *Trichoderma* spp. includes: *Armillaria, Botrytis, Chondrostereum, Colletotrichum, Dematophora, Diaporthe, Endothia, Fulvia, Fusarium, Fusicladium, Helminthosporium, Macrophomina, Monilia, Nectria, Phoma, Phytophthora, Plasmopara, Pseudoperonospara, Pythium, Rhizoctonia, Rhizopus, Sclerotinia, Sclerotium, Venturia, Verticillium* and fungi that cause wood rot (Monte, 2001).

18

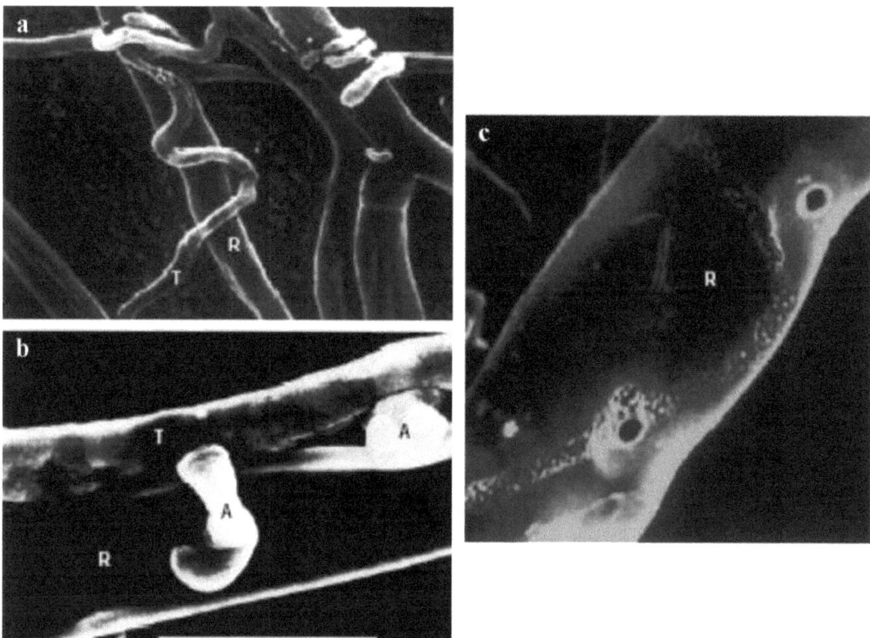

Figura 6. Process of mycoparasitism of *Trichoderma* sp. on *Rhizoctonia solani.* (a) formation of the appressorium; (b) parasitism on the host hyphae; (c) holes left by the appressorium.

T- *Trichoderma* spp. R- *Rhizoctonia solani* A- appressoria (adapted from Harman *et al.* 2004).

3. MATERIAL AND METHODS

3.1 Microorganisms

The microorganisms used were isolates of *Fusarium sp.* and *Trichoderma* sp. according to tables 1 and 2.

Table 1. Isolates of *Fusarium* sp. used in this study.

Isolate/Strain	Geographical Origin	Race	VCG
589	Rio Grande do Sul	NI	NI
859	Rio Grande do Sul	NI	NI
921	Rio Grande do Sul	NI	NI
921/2	Rio Grande do Sul	NI	NI
1205/1	Rio Grande do Sul	NI	NI
1205/2	Rio Grande do Sul	NI	NI
Fusarium 23°	Holy Spirit	2	NI
Fusarium 27°	Holy Spirit	1	NI
TO 11°	Sao Paulo	2	NI
TO 245°°	Sao Paulo	2	NI
26380*	United States	3	0033
34970*	United States	2	0030
MM 66**	United States	2	0032
OSU 451**	United States	2	0031

° Granted by Dr. Ailton Reis of Embrapa Hortaliças, Federal District.

°° Granted by Dr. Rômulo Kobori of Sakata Seeds through Dr. Wagner Bettiol of Embrapa Meio Ambiente, Jagariûna-SP.

* Granted by Dr. Kerry O' Donnell of the United States Department of Agriculture.

** Granted by Dr. Liane Rosewich Gale of the Department of Plant Pathology at the University of Minnesota.

NI- not identified.

The isolates Fusarium 23, Fusarium 27, TO 11 and TO 245 had their races identified and will serve as a control for their identification in the pathogenicity tests.

The strains 26380, 34970, OSU 451 and MM66 were previously identified as Fol, and are the standard strains of the vegetative compatibility groups of this special form.

The isolates 589, 859, 921, 921/2, 1205/1 and 1205/2 of Fusarium sp. belong to the collection of the Biological Control of Plant Diseases laboratory and had no identification of species, race or vegetative compatibility group.

Table 2. Isolates of *Trichoderma* sp. used in this work.

Isolated	Culture of origin	Geographical origin
T1	Vine	Rio Grande do Sul
T2	Gerbera	Rio Grande do Sul
T3	Cucumber	Rio Grande do Sul
T4	Vine	Rio Grande do Sul
T6	Perfect love	Rio Grande do Sul
T8	-	Rio Grande do Sul
T15	Apple tree	Rio Grande do Sul
T17	Apple tree	Rio Grande do Sul
T19	Apple tree	Rio Grande do Sul
T5A	-	-

3.2 Culture Media and Solutions

3.2.1 BDA medium

BDA (Potato Dextrose Agar) medium was used to maintain, grow and evaluate antagonism. 200 grams of peeled potato were weighed and boiled in 1 liter of distilled water for 20 minutes, the broth was filtered and 20 grams of glucose, 20 grams of agar and autoclaved water were added to the resulting filtrate until it was 1L, after which it was autoclaved for 20 minutes at 121 °C and 1 atm of pressure.

3.2.2 Saline solution

It was made up of 9 grams of NaCl in 1 L of distilled water and then autoclaved for 20 minutes at 121 °C and 1 atm of pressure. The flasks were kept at room temperature.

3.2.3 Basal medium

The following reagents were added to 1 L of distilled water: 30 g of sucrose, 1 g of KH_2PO_4, 0.5 g of $MgSO_4.7\ H_2O$, 0.5 g of KCl, 10 mg of $FeSO_4.7\ H_2O$, 20 g of agar and 0.2 mL of a trace element solution and autoclaved at 121 °C and 1 atm for 15 minutes.

3.2.4 Trace element solution

In 95 mL of distilled water were added: 5g citric acid, 5g $ZnSO_4 . 7 H_2 O$, 1g $Fe(NH_4)_2 (SO_4)_2 . 6H_2O$, 1g $CuSO_4. 5 H_2O$, 1g $Mn SO_4 . H_2O$, 50 mg H_3BO_3, 50 mg $NaMoO_4 . 2 H_2O$ and autoclaved at 121 °C and 1 atm for 15 minutes.

3.2.5 Minimum means

In 1 L of basal medium, 2 g of NaNO3 was added and autoclaved at 121 °C and 1 atm for 15 minutes.

3.2.6 Potato dextrose agar medium with chlorate

The potato dextrose agar medium was made as mentioned in section 3.2.1, then 15 g of KClO3 was added and autoclaved at 121 °C and 1 atm for 15 minutes.

3.2.7 Maintenance of microorganisms

The microorganisms were kept in BDA medium and repicked every 4 weeks. After growing in a greenhouse for 7 days, the repiques were kept refrigerated at 4° C until a later repique.

3.2.8 Suspension of conidia

For the production of *Trichoderma* spp. conidia suspensions, plates containing BDA medium were used, on the surface of which the strains were grown for 7 days at 28 °C with 12-hour photophase in a BOD incubator. Sterilized distilled water was added to the plates and the surface was washed using a Drigalski loop. Enough autoclaved distilled water was added to complete the 50 mL suspension. The concentration of conidia in the suspension was determined using a Neubauer chamber.

3.2.9 Production of conidia of *Fusarium* sp. and *Trichoderma* sp.

Moistened and sterilized crushed popcorn was used as a growth substrate for the fungus. The primary inoculum of the fungus was mixed with 100 g of grain in plastic bags. After seven days of colonization at a temperature of 25°C and a photoperiod of 12 hours, the corn was incorporated into the rooting substrate of the plants, in the proportion of 20 g/kg of substrate according to Melo & Valarini (1995). The final concentration of conidia was determined in a Neubauer chamber from a suspension made with 1g of colonized substrate diluted in 10 mL of saline solution, and viability was assessed by dilution plating on BDA medium and counting colonies.

For the production of *Trichoderma* sp. conidia, the same technique was used, replacing popcorn with rice.

3.3 *Fusarium* pathogenicity test *and* identification of races

A suspension was made from maintenance plates of *Fusarium* sp. isolates by adding distilled and autoclaved water and washing the surface of the medium with a Drigalski loop to release the mycelium and conidia. Afterwards, the concentration of the suspension was adjusted to 1 x 10^6 propagules per mL, by dilution when necessary.

The following set of cultivars were used to differentiate between races of *Fusarium oxysporum* f. sp. *lycopersici*: Super Marmande (susceptible to all races), Industrial (resistant to race 1), Rasteiro (resistant to races 1 and 2), produced by Isla seeds. The seeds of the cultivar BHRS-2.3 (resistant to all three races of the pathogen) were supplied by Dr. Ailton Reis, Embrapa Hortaliças and the experiments were carried out in accordance with Reis *et al.*, (2004).

The seeds of the genotypes were sown in the commercial Plantmax® autoclaved substrate, in plastic trays measuring 29 cm x 43 cm, kept in a greenhouse for 14 days for rooting. After this period, the seedlings were removed from the trays and their roots washed in running water to remove the substrate. Next, a segment of about 0.5 cm from the end of each root was cut off and discarded. The root system of each seedling was then immersed in the pathogen spore suspension for 5 minutes and then transferred to 128-cell Styrofoam trays filled with autoclaved commercial substrate. For the control treatment, only distilled and autoclaved water was used. For each of the tests, 1 mL of the suspension of each fungus was also applied to the cells in the tray directly onto the rooting substrate. Each treatment was evaluated on 10 plants, with two replicates.

Irrigation was carried out daily with drinking water, increasing the amount according to the development of the seedlings.

The severity of the disease was assessed at 15 and 30 days, based on a scale of scores ranging from 0 to 3 according to Vakalounakis *et al.* (2004), related to the presence of visible symptoms on the aerial part, such as the appearance of leaves with a burnt, brown appearance at the base of the plants; twisted, chlorotic and/or wilted leaves; and dead plants. The following scale was used according to the percentage of visible symptoms on the aerial part: (0) - no symptoms;

(1) - mild or moderate wilting, slight vascular discoloration of the stem;

(2) - severe wilting and vascular discoloration;

(3) - dead plant.

To confirm death by fusariosis, the plants that showed signs of the disease were treated in a 70% alcohol solution for 1 minute, then immersed in a 10% hypochlorite solution for the same period of time, left to dry in a chapel and placed in Petri dishes containing BDA medium according to Alfenas & Mafia (2007). Confirmation of death by *Fol* was made by observing the growth of colonies of the phytopathogen from the plants and the growth and reproduction structures under an optical microscope.

3.4 Antagonism tests

The antagonism of *Trichoderma* sp. isolates against *Fol* isolates was assessed in two ways: using the Bell *et al.* scale (1982) to assess mycoparasitism in relation to the phytopathogenic fungus, and using the methodology described by Abdell-Fattah *et al.* (2007).

The isolates of *Fusarium* sp. that proved to be pathogenic to tomatoes in the pathogenicity tests were used for antagonism. Also included in the tests were strains 26380 and 34970 of *F. oxysporum* f. sp. *lycopersici*, standard strains from the vegetative compatibility groups and belonging to race 3.

3.4.1 Evaluation according to the Bell *et al.* scale (1984)

An agar disk colonized by the phytopathogenic fungus was transferred to Petri dishes containing BDA culture medium, using a Zeni tube, 0.5 cm from the edge of the dish, which was then kept in a BOD oven for 48 hours. After this period, a disc of agar colonized by an isolate of *Trichoderma* sp. was transferred to the same plate 0.5 cm from the edge of the plate at a point equidistant from the inoculum of the *Fusarium* sp. isolate, as shown in Figure 7. The plates were kept in an oven at 26° C (± 2°C) with 12-hour photophase. After eight days, the growth of the fungi was evaluated, based on the adapted criteria of Bell *et al.* (1982), which used the following scale of scores from 1 to 5 (Figures 7 and 8): (1) antagonist grows and occupies the entire plate; (2) antagonist grows, overlapping the pathogen colony (2/3 of the plate); (2.5) antagonist grows, overlapping only the edge of the pathogen colony; (3) antagonist and pathogen grow up to half the plate (no overlap); (4) pathogen grows overlapping the antagonist's colony (2/3 of the plate); (5) pathogen grows and occupies the entire plate.

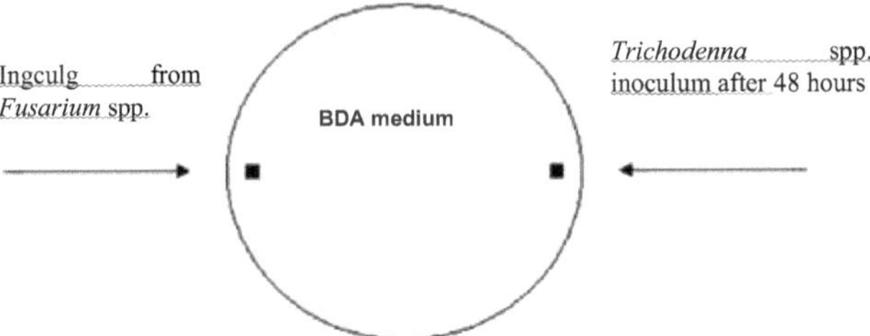

Figure 7. Schematic drawing of the direct confrontation test between isolates of *Fusarium* sp. and *Trichoderma* sp.

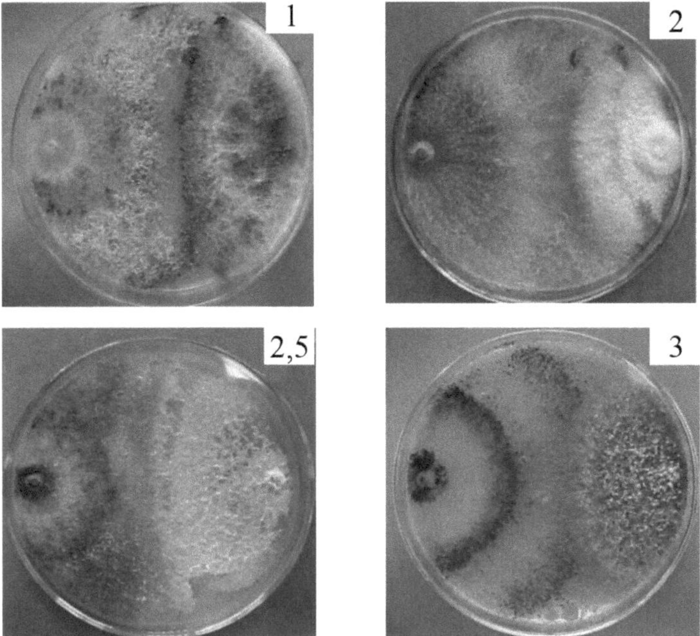

Figure 8. Representation of the modified Bell *et al.* (1982) grading scale used in the direct confrontation tests to evaluate antagonism on the plates (1) Antagonist grows and occupies the entire plate; (2) Antagonist grows, overlapping the pathogen colony (2/3 of the plate); (2.5) Antagonist grows, overlapping only the edge of the pathogen colony; (3) Antagonist and pathogen grow up to half the plate).

3.4.2 Evaluation of *Fusarium* sp. colony growth inhibition

The growth of the Fusarium sp. colonies was measured 8 days after the *Trichoderma* sp. inoculum was added to the direct confrontation tests. Using a caliper, the radius of the colony of the phytopathogenic fungus confronted with the isolates of the antagonist was measured, as well as the control group containing only the isolate of *Fusarium sp.*

The results were then evaluated according to the following formula from Abdell-Fattah *et al.* (2007) to calculate the percentage of growth inhibition caused by the *Trichoderma* sp. isolates:

% inhibition = [control growth - test growth / control growth] X 100; which indicates the percentage of inhibition of the antagonized fungus colony in relation to the control.

3.5 Volatile metabolite production test

The effect of the volatile metabolites produced by *Trichoderma* sp. *in vitro* was evaluated according to the methodology described by Dennis & Webster (1971b) and Figure 9. Two bottoms of Petri dishes containing BDA medium were individually inoculated using a Zeni tube, from colonies of the pathogen and the antagonist, grown on control plates for five days, the bottoms being adjusted and closed with crepe tape. The control did not contain the antagonist *Trichoderma* sp. The cultures were incubated at 28°C± 0.5°C, with 12-hour photophase. The treatments were conducted with three replicates. Radial growth was measured after 48 and 120 hours of incubation.

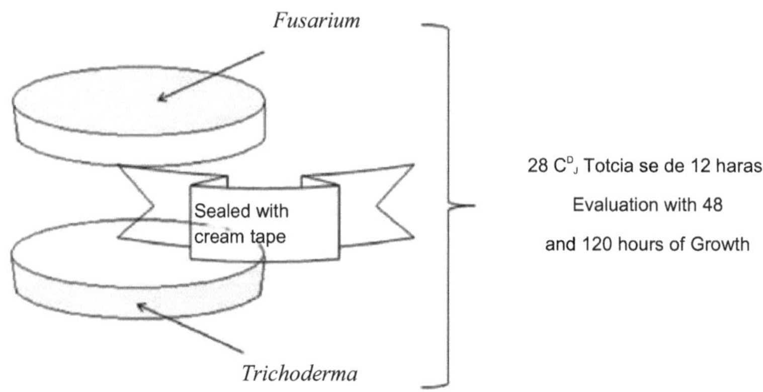

Figure 9. Schematic drawing of the test for the production of volatile metabolites by *Trichoderma* sp.

3.6 *Trichoderma* **biological control tests on** *Fusarium* **f. sp.** *lycopersici*

This experiment was carried out on tomato plants as a preventative measure in accordance with Melo & Valarini (1995). Isolates T17 and T6 of *Trichoderma* sp. were used, determined according to the direct confrontation experiment between *Trichoderma* sp. and *Fusarium sp. Fol* isolates 1205/2 and TO 245 were used, selected on the basis of the results obtained in the pathogenicity tests.

Ten days before sowing the tomato seeds, the soil was treated with 20 g of crushed maize colonized by Fusarium sp./kg of soil. After this period, the population of Fol present in one gram of soil mixed with 10 mL of saline solution was determined in a Neubauer chamber. The viability of *Fusarium* sp. propagules was checked by determining the number of CFU present in the substrate on BDA medium. For the biological treatment, rice colonized by isolates T6 and T17 of *Trichoderma* sp. was washed in water until a suspension of 1 x 10^8 conidia/mL was obtained. Tomato seedlings approximately 15 days old had the ends of their roots cut off (0.5 cm) and were then immersed in one of the *Trichoderma* sp. suspensions for a period of 5 minutes, depending on the treatment. The seedlings were then transplanted into substrates infested with one of the *Fol.* 1205/2 or TO 245 isolates. Each treatment was carried out on 10 plants, accompanied by the control treatment.

Plant death was measured weekly and, after 45 days, the presence or absence of disease symptoms was assessed according to the disease severity scale described by Vakalounakis *et al.* (2004). Plant height and root length were measured using a millimeter ruler. Dry weight was determined by separating the aerial part from the roots and drying each part in an oven for around 48 hours at 50 °C. Confirmation of death was carried out according to Alfenas & Mafia (2007), by identifying the reproductive structures of the fungus under an optical microscope.

3.7 Analysis of vegetative compatibility groups (GCV)

3.7.1 Generation of non-nitrate-utilizing mutants

The tests were carried out in accordance with Correl *et al.* (1987), with isolates of *Fusarium* sp. being repotted from a control plate onto plates containing BDA medium plus chlorate. After repotting, the inoculated plates were kept in a BOD oven at 28 °C with 12-hour photophase. Sectors with the fastest growth were analyzed daily and repotted onto minimal medium containing nitrate as the sole source of nitrogen. In this medium, colonies that grew sparsely and without aerial mycelium were considered *nit.* mutants.

3.7.2 Characterization of *nitrogen* mutant phenotypes

The phenotypes of these mutants were assigned to different phenotypic classes based on their growth on medium containing one of the following different nitrogen sources: minimal medium, nitrite medium and hypoxanthine medium. For identification, the mutant colonies were repotted onto each of these media and their growth compared with the parental one after 4 days of growth according to Table 3.

Table 3. Identification of non-nitrate-utilizing (*nit*) mutants of *Fusarium oxysporum* by growth on different nitrogen sources (adapted from Correl *et al.* 1987).

Change	Designation of the mutant	Growth on different nitrogen sources [a]		
		Nitrate	Nitrite	Hypoxanthine
None	Wild	+	+	+
Nitrate reductase (structural locus)	*nit* 1	-	+	+
Co-factor Molybdenum (*Loci*)	NitM	-	+	-

(a) Growth in basal medium with different nitrogen sources; (+) typical wild growth; (-) sparse growth without aerial mycelium.

4.7.3 Complementary tests

Vegetatively compatible mutants must complement each other on minimal medium, with the development of dense aerial mycelium where the colonies of the two *nit* mutants meet. The pairing was done by removing the mycelium from each colony using a Zeni tube and placing it about 1-2 cm apart in a Petri dish containing minimal medium. The plates were incubated as described above for about 14 days and then evaluated for complementation (occurrence of hyphae anastomosis and heterochorium formation). All *nit* mutants recovered from the same parent were paired with at least one *nit1* and one NitM from the same parent. A nit1 mutant and a nitM mutant from each isolate were paired in all possible combinations with a nit1 mutant and a nitM mutant from all isolates to confirm the GCVs.

3.8 Statistical methods

The results were evaluated by One-Way ANOVA analysis of variance, with Tukey's post-test for $p < 0.05$, using the GraphPad Prism 5.0 computer program.

4. RESULTS AND DISCUSSION

4.1 Pathogenicity tests

Although they came from tomato plants, for most of the isolates from Serra Gaúcha, after 15 days of testing no disease symptoms were observed on any of the cultivars used. Only isolate 1205/2 from *Fol* caused disease symptoms such as yellowing and necrosis of the older leaves, wilting of the plant followed often by death, in the susceptible cultivars and in the cultivar resistant to race 1, reaching level 3 on the scale, with death of all the contaminated seedlings and no signs of disease being observed in the cultivar resistant to races 1 and 2, as well as for the cultivar resistant to race 3 (Figure 10).

This avirulence observed for the isolates from the Serra Gaúcha may be due to the long freezer storage time of some samples. Another hypothesis could be the pressure of selection for avirulence due to the frequent repotting of these isolates in the culture medium.

Figura 10. Tomato seedlings inoculated with isolate 1205/2 after 3 days of treatment. A- susceptible cultivar with disease symptoms. B- cultivar resistant to races 1 and 2 without symptoms.

The TO 11 and TO 245 isolates from Sao Paulo caused disease symptoms on susceptible and resistant cultivars to race 1, with scores of 1.7 and 2.6 according to the severity scale of Vakalounakis *et al.* (2004). However, no disease symptoms were observed on cultivars resistant to races 2 and 3.

With the isolates Fusarium 23 and 27, from Espirito Santo, symptoms were observed on the susceptible cultivar. In the cultivar resistant to race 1, however, only Fusarium 23 caused symptoms of fusariosis on the tomato seedlings. No disease symptoms were observed on the other cultivars.

After 30 days of evaluation of the pathogenicity results, according to the scale used, the highest severity was observed for isolate 1205/2, reaching level 3 in the susceptible cultivar and 2.8 in the cultivar resistant to the race. This isolate was classified as belonging to race 2 since no disease symptoms were observed on the cultivar resistant to races 1 and 2, nor on the cultivar resistant to race 3.

Isolate TO 245 scored 2.6 and isolate TO 11 scored 2.55 on a cultivar resistant to race 1 (Table 4).

For the isolate Fusarium 27, there was no incidence of disease in the cultivar resistant to race 1, and for Fusarium 23, the disease severity score was 1.70 (Table 4). These results confirm the classification into races 1 and 2 respectively, in line with the results obtained by Reis *et al.* (2005).

None of the isolates from *Fol* da Serra Gaúcha caused disease symptoms on the cultivar resistant to races 1 and 2, ruling out the possibility of any of the isolates belonging to race *3*.

The other isolates tested did not cause disease in any of the cultivars treated, and were classified as possible avirulent. Similarly, Cai *et al.* (2003) obtained three isolates of *Fol* from plants with disease symptoms that did not cause symptoms on susceptible cultivars, being classified as endophytic and non-pathogenic.

No isolate belonging to race 3 was detected among the six isolates of *Fusarium* sp. obtained from plants with symptoms of fusarium wilt in the state of Rio Grande do Sul.

The isolates TO 11, TO 245, Fusarium 23 and Fusarium 27 were used as a control to identify the races in the cultivars, since they had a known race and were confirmed in the pathogenicity tests carried out.

Table 4 - Severity of disease caused by isolates of *Fusarium* sp. after 15 and 30 days of experimentation on tomato cultivars susceptible and resistant to race 1, races 1 and 2 and race 3 according to the scale of Vakalounakis *et al.* (2004): (0) - no symptoms; (1) - mild or moderate wilting, slight vascular discoloration of the stem; (2) - severe wilting and vascular discoloration; (3) - dead plant.

Isolated from	Tomato cultivars[a]							
Fusarium	Super Marmande		Industrial		Rast teiro		BHRS-2,3	
	15 days	30 days	15 days	30 days	15 days	30 days	15 days	30 days
589	0	0	0	0	0	0	0	0
859	0	0	0	0	0	0	0	0
921	0	0	0	0	0	0	0	0
921/2	0	0	0	0	0	0	0	0
1205/1	0	0	0	0	0	0	0	0
1205/2	3	3[a]	2,8[a]	2,8[a]	0	0	0	0
Fusarium 23	0,9[c]	1,7[b]	1,7[a]	1,7[a]	0	0	0	0
Fusarium 27	1,9[ab]	2,2[ab]	0	0	0	0	0	0

TO 11	$1,5^b$	$2,7^{ab}$	$1,7^a$	$2,55^a$	0	0	0	0
TO 245	$1,8^{ab}$	$2,4^{ab}$	$2,6^a$	$2,6^a$	0	0	0	0
Control	0	0	0	0	0	0	0	0

* Averages followed by the same letter on each evaluation day do not differ according to Tukey's test (p< 0.05).

[a]- Differentiating cultivars used for breed designation: Super Marmande (susceptible), Industrial (resistant to breed 1), Rasteiro (resistant to breeds 1 and 2) and BHRS-2.3 (resistant to breeds 1, 2 and 3).

The re-isolation of the causal agent of the disease confirmed the death caused by fusariosis, as colonies of *Fusarium* sp. could be seen on the plates of all the dead plant samples (Figure 11).

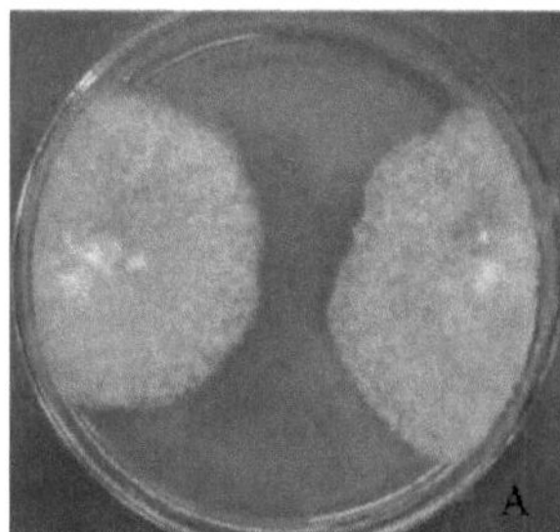
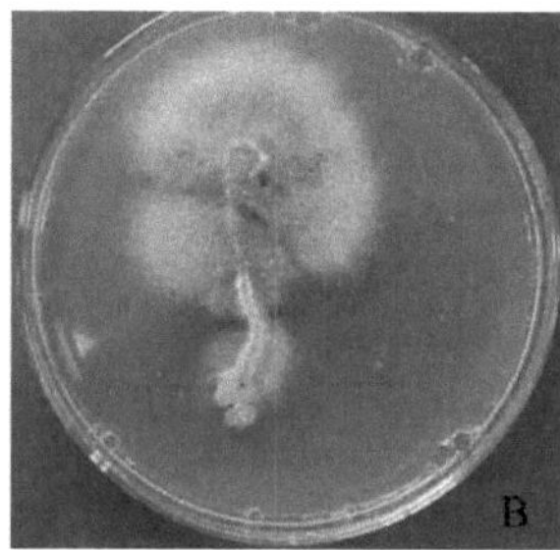

Figura 11. Isolates of *Fusarium* sp. obtained from dead plants treated with isolates 1205/2 (A) and TO 11 (B)

4.2 *In vitro* antagonism of *Trichoderma* against *Fusarium* f. sp. *lycopersici*

4.2.1 Results according to the Bell Scale

Table 5 shows the results obtained, and it can be seen that the lowest overall averages on the scale were achieved with isolates T3, T8 and T17. For the Fusarium 23 isolate, the statistical difference between the values of the effects achieved by each of the *Trichoderma* sp. isolates confronted was not significantly different, but they all differed from the control treatment containing only the *Fol* isolate.

For the isolate Fusarium 27, the lowest averages were presented with isolates T3, T8 and T17 of *Trichoderma* sp. with evidence of mycoparasitism on the phytopathogenic fungus. On isolate TO 11 of Fol, isolates T8 and T17 of *Trichoderma* sp. occupied 2/3 of the surface of the culture medium, sporulating and overlapping the host colony, suggesting a strong mycoparasitic action (Figure 12). Similarly, isolates T1, T3, T8 and T 17 grew over isolate TO 245, with T17 completely overlapping the pathogen's colony.

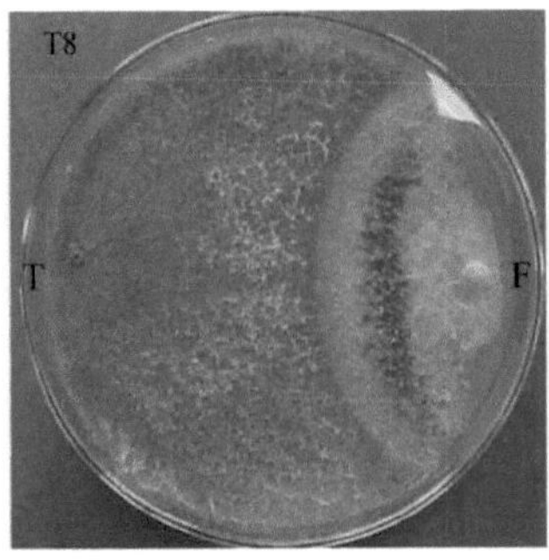

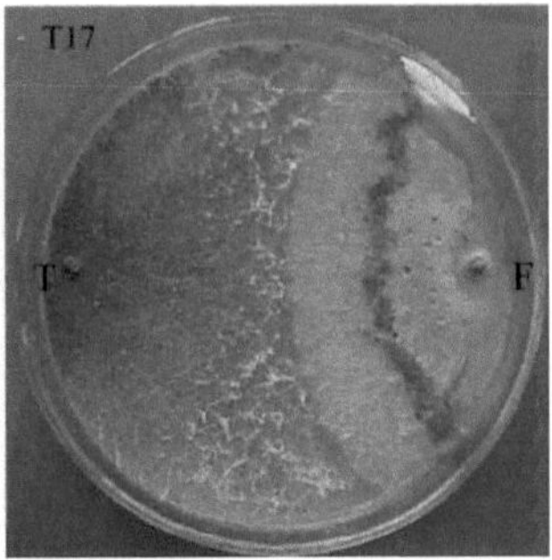

Figura 12. Mycoparasitism of isolates T8 and T17 against isolate TO 11. On each plate, (T) = *Trichoderma*, (F) = *Fusarium*.

When evaluating the antagonism of all the *Trichoderma* sp. isolates against *Fusarium* sp. strain 26380, all the results were significant in relation to the control treatment. The greatest difference was found in direct comparison with isolate T15. Isolates T1, T3, T17, T19 and T5A did not differ from isolate T15 of *Trichoderma sp.*, with evidence of overlapping of the colony of the phytopathogenic fungus in at least one of the replicates tested (Figure 13).

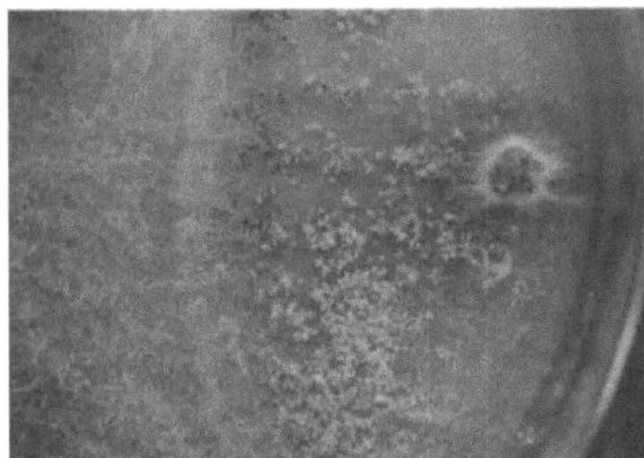

Figura 13.   Colony overlap of Fol strain 26380 by *Trichoderma* sp. isolate T1.

With regard to *Fusarium* sp. strain 34970 from the United States, total overlap of the phytopathogen colony was only observed with isolate T17. There was no statistical difference between the other isolates, but they all differed from the control treatment.

Mycoparasitism was only observed for isolate 1205/2 of *Fusarium* sp. from the Serra Gaúcha against isolates T3 and T8 of *Trichoderma* sp. The other antagonist isolates did not differ significantly from the control treatment.

Table 5. *In vitro* antagonism scores between *Trichoderma* sp. strains and isolates of *Fusarium* sp. according to the scale of Bell *et al.* (1982) (1)- Antagonist grows and occupies the whole plate; (2)- Antagonist grows, overlapping the pathogen colony (2/3 of the plate); (2.5)- Antagonist grows, overlapping only the edge of the pathogen colony; (3)- Antagonist and pathogen grow up to half the plate (neither organism overlaps the other); (4)- Pathogen grows overlapping the antagonist colony (2/3 of the plate); (5)- Pathogen grows and occupies the entire plate;

Isolated from Trichoderma	Isolates of *F. oxysporum* f. sp. *lycopersici* Notes on antagonism								
	Fus. 23	Fus. 27	TO 11	TO 245	26380	34970	MM66	OSU 451	1205/2
T1	3^{bA}	$_3$bA	$_3$bA	$_2$cA	cdA	2,83^{bA}	2,33^{bA}	2,83^{bA}	$_3$aA
T2	3^{bA}	$_3$bA	$_3$bA	$_3$bA	$_3$bA	$_3$bA	$_3$bA	$_3$bA	$_3$aA
T3	3 B^{b}	2,33^{cAB}	$_3$bB	cAB	$_{2,66}$bcdAB	$_3$bB	2,5^{bAB}	2,5^{bAB}	1,66^{bA}
T4	3^{bA}	$_3$bA	$_3$bA	$_3$bA	2,83^{bcA}	$_3$bA	$_3$bA	$_3$bA	$_3$aA
T6	3^{bA}	$_3$bA	$_3$bA	$_3$bA	$_3$bA	$_3$bA	$_3$bA	$_3$bA	$_3$aA
T8	3^{bA}	bcA	$_2$cA	$_2$cA	$_3$bA	$_3$bA	1,66^{cA}	2,83^{bA}	2,66^{abA}
T15	2.66 A^{b}	$_3$bA	$_3$bA	$_3$bA	1,66^{dA}	2,5^{bA}	2,83^{bA}	2,5^{bA}	$_3$aA
T17	2.66 A^{b}	$_2$cA	$_2$cA	1,66^{cA}	2,16^{cdA}	1,83^{cA}	$_2$cA	1,66^{cA}	$_3$aA
T19	3^{bA}	$_3$bA	$_3$bA	2,66^{bA}	2,33^{cdA}	2,83^{bA}	2,66^{bA}	2,5^{bA}	$_3$aA
T5A	3^{bA}	$_3$bA	$_3$bA	$_3$bA	2,33^{cdB}	2,66^{bAB}	$_3$bA	$_3$bA	$_3$aA
Witness	4^{a}	$_4$a	4,66^{a}	$_4$a	$_4$a	$_4$a	$_4$a	$_4$a	$_4$a

*Averages followed by the same small letter in the columns do not differ according to Tukey's test (p< 0.05).
** Averages followed by the same capital letter in the rows do not differ by Tukey's test (p< 0.05). Fus. 23= Fusarium 23; Fus. 27= Fusarium 27.

The results obtained indicate the potential of the *Trichoderma* sp. isolates in relation to their mycoparasitism capacity when compared to the results observed by Gómez *et al.* (1997), in which ten *Trichoderma sp.* isolates showed greater antagonism capacity against *Rhizoctonia* sp., with only one overlap result in a direct confrontation test against *Fusarium* sp. The results found in the direct confrontation tests in the current study suggest the potential for applying isolates T3, T8 and T17 in the control of *Fol,* since they showed overlapping colonies in at least six of the nine isolates of *Fusarium sp.* confronted.

Biocontrol capacity, however, depends on several factors in addition to mycoparasitism, and the direct confrontation method is one of the ways of selecting isolates with potential for application in biological control. This hypothesis can be reinforced by the work of Sànchez *et al.* (2007) who observed that the parasitism of *T. longibrachiatum* on *Thielaviopsis paradoxa* is done through the production of extracellular enzymes that degrade the constituents of the cell wall and non-volatile diffusible metabolites that are also involved in antagonism by *Trichoderma* sp.

4.2.2 Inhibition of colony growth of *Fusarium oxysporum* **f.sp.** *lycopersici*

The average growth inhibition data ranged from 31% to 53%. The T3 isolate of *Trichoderma* sp. inhibited the fungal development of the phytopathogen Fusarium 23 by an average of 53%, while there was no statistical difference between all the other isolates of the antagonist, except the T19 isolate (Table 6).

The behavior of the Fusarium 27 isolate in relation to the *Trichoderma* sp. isolates served as a basis for dividing the latter into two groups: (1) they caused greater inhibition of the target fungus, such as isolates T2, T3, T4, T6 and T8; and (2) they caused less inhibition, represented by the remaining isolates. The highest value was 47% inhibition and the lowest 37%.

The TO 11 isolate of *Fusarium* sp. showed the highest inhibition averages, with the highest value of 53%, determined by isolates T1, T2 and T4 (Figure 14), and the lowest value of 42% against *Trichoderma* sp. isolate T15.

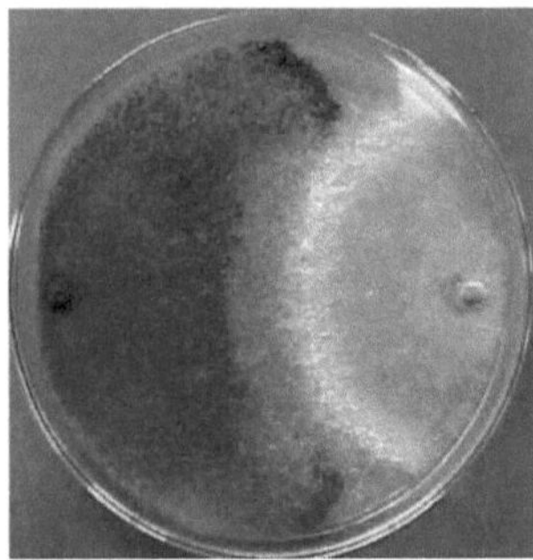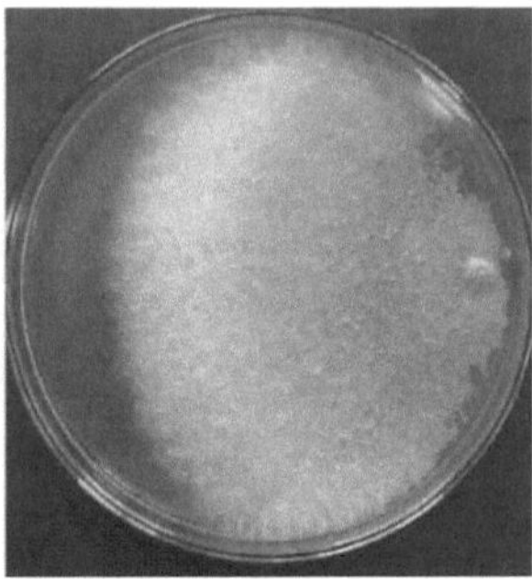

Figure 14. Growth inhibition of isolate TO 11 against isolate T4.

There was no significant difference between the isolates of *Trichoderma* sp. against isolate TO 245, with an average total inhibition value of 36%.

For *Fol* strain 26380, the greatest inhibition was when directly confronted with isolates T2 or T3, with 44% and 47%, respectively, **and** for pathogen strain 34970 there were no significant growth inhibition results.

The direct comparison tests of isolates T3 and T8 of *Trichoderma* sp. against isolate 1205/2 of *Fusarium* sp. showed the highest average inhibition (46%). The lowest inhibition (30%) was found when compared against *Trichoderma sp. isolate* T5A (Table 6).

The MM6 strain showed the lowest average inhibition values against *Trichoderma* sp. isolate T19, with only 29%, and the highest average inhibition values against isolate T1, with 44% inhibition.

For *Fol* strain OSU 451, *the* lowest inhibition values were observed against isolates T19, T15 and T2, and the highest inhibition against T1, T4, T6 and T5A.

Table 6. Percentage of growth inhibition of *Fusarium* isolates caused by *Trichoderma* in direct confrontation test, according to the equation by Abdell-Fattah *et al.* (2007).

Isolated from *Trichoderma*	Isolates of *F. oxysporum* f. sp. *lycopersici* Percentage inhibition								
	Fus. 23	Fus. 27	TO 11	TO 245	26380	34970	MM 66	OSU 451	1205/2
T1	^abAB	41bB	53aA	44aAB	abB	35aB	44abB	abAB	abB
T2	^yabAB	aAB	53aA	41aB	44aAB	31aC	bcdC	34dC	abAB
T3	53aA	43aB	49abAB	38aBC	aAB	34aC	bcdBC	abcdBC	46aAB
T4	48abAB	45aAB	53aA	41aBC	abBC	36aBC	bcdC	abBC	abBC
T6	^yabA	aAB	51abA	38aBC	38abBC	33aC	bcdBC	44aA	40abBC
T8	48 A ab	aAB	51abA	aAB	35bB	35aB	46aAB	abcdAB	46aAB
T15	46 A ab	37bBC	bAB	34aBC	40abA	31aC	bcdC	34dBC	35abBC
T17	^yabAB	37bB	48abA	aAB	abAB	aAB	abcAB	abcdB	abB
T19	41bA	37bB	46abA	36aBC	abB	30aC	29dC	33dBC	35abB
T5A	44 A ab	36bAB	46abA	aAB	40abAB	31aB	bcdAB	41abA	30bB

*Averages followed by the same small letter in the columns do not differ according to Tukey's test ($p < 0.05$).

** Averages followed by the same capital letter in the rows do not differ by Tukey's test ($p < 0.05$). Fus. 23= Fusarium 23; Fus. 27= Fusarium 27.

In addition, all the *Trichoderma* sp. isolates grew faster than the *Fusarium* sp. isolates, occupying around half of the plate even when replanted 48 hours after *Fol.* Similar results were found by Abdel-Fattah *et al.* (2007), who recorded a 48% inhibition in the growth of *Bipolaris oryzae* colonies when confronted with *Trichoderma* sp. The same speed of growth of the antagonist was observed after eight days of testing, which is one of the advantages of using isolates of this genus.

Comparing the two ways of evaluating the direct confrontation tests, it can be seen that isolates such as T2, which showed results of little overlap between colonies of the pathogen, with an average of 3.00 on the scale, showed high growth inhibition averages, with the highest inhibition rates for isolates TO 11 and Fusarium 27. Isolate T3, which showed overlap against six of the nine *Fol isolates,* also showed high inhibition averages. The same antagonism results were also observed for T8 and T17 in relation to the *Fol isolates.*

However, these results do not allow us to conclude on the relationship between mycoparasitism and the inhibitory effect of *Trichoderma* sp. There are no results to indicate which of the evaluation methods used is the most important for the biocontrol capacity of species belonging to this genus. Or, if this ability results from the synergistic action of all the mechanisms, which methods would be the most suitable for selecting isolates in the laboratory.

4.3 Evaluation of the production of volatile metabolites by *Trichoderma* sp.

Growth inhibition, through the production of volatile metabolites, was tested with three isolates of

Trichoderma sp. (T3, T8 and T17), selected in antagonism tests in direct confrontation against all isolates of *F. oxysporum* f. sp. *lycopersici, using the* highest incidence of mycoparasitism as a criterion.

According to Figure 15, growth inhibition of the Fusarium 23 isolate by *Trichoderma* sp. isolate T17 was observed late in the 120 hours, with a 0.83 cm reduction in colony diameter.

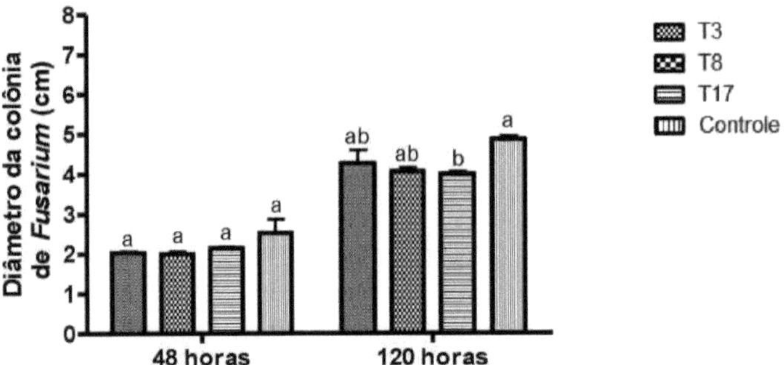

Figura 15. Colony diameter of *Fusarium* sp. isolate 23 in an inhibition test using volatile metabolites released by *Trichoderma* sp. compared to the control.

* Averages followed by the same letter for each time evaluated do not differ according to Tukey's test ($p <$ 0.05).

Caption:

In relation to the Fusarium 27 isolate, the inhibition induced by all the *Trichoderma* isolates was significant at 120 hours (Figure 16). The average measurements of the diameter of the colonies after 120 hours of growth can be seen. In the control group, the average was 6.76 cm and 3.75, 3.8 and 3.8 cm for T3, T8 and T17, respectively.

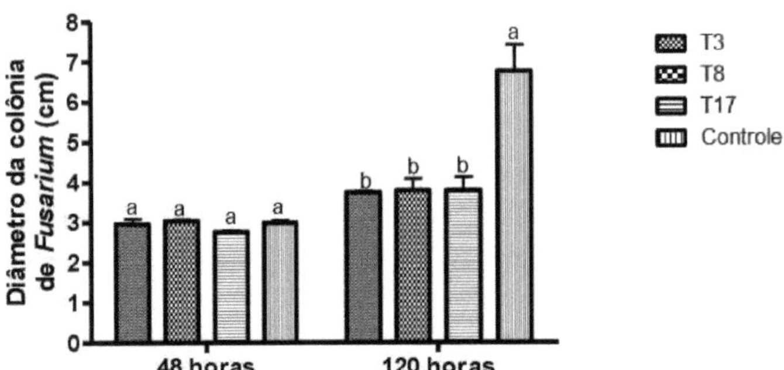

Figura 16. Colony diameter of *Fusarium* sp. isolate 27 in an inhibition test by volatile metabolites released by *Trichoderma* sp. isolates in relation to the control.

* Averages followed by the same letter for each time evaluated do not differ according to Tukey's test ($p <$

0.05).

Figure 17 shows that in addition to the reduction in colony diameter, the metabolites released by the T3 isolate of *Trichoderma* sp. cause morphological changes in which the hyphae of the Fusarium 27 isolate appear thinner. Similarly, Dal Bello *et al.* (1997) also observed that the effect of *Trichoderma* sp. volatile metabolites can manifest itself directly on the hyphae of phytopathogens (thinning and darkening) and by preventing the formation of sclerotia, without necessarily affecting growth and sporulation, but by producing an indirect inhibitory effect.

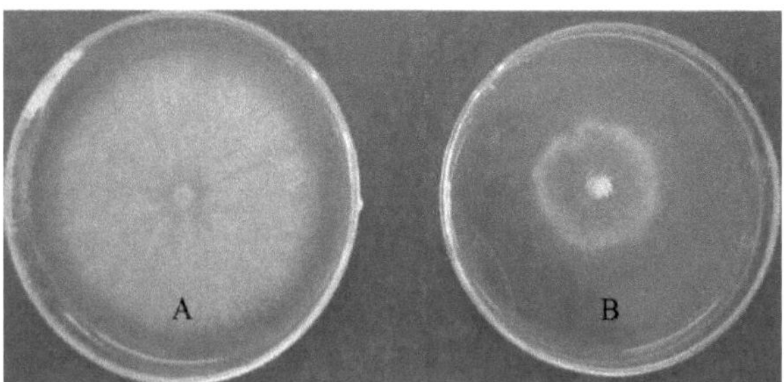

Figure 17. Comparison between colonies of Fusarium isolate 27, control group (A), and in the presence of volatile metabolites released by *Trichoderma* sp. isolate T3 (B).

Reinforcing the data reported by Paradela (2002), that *Fusarium* sp. isolates can respond differently to the antagonistic actions of microorganisms, the results shown in Figure 18 demonstrate the high resistance of isolate TO 11, which was not inhibited by the volatile metabolites released by the *Trichoderma* sp. isolates evaluated.

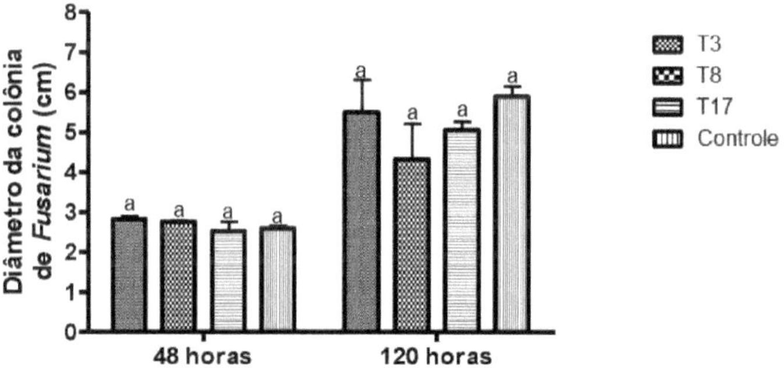

Figura 18. Colony diameter of *Fusarium* sp. isolate TO 11 in an inhibition test using volatile metabolites released by *Trichoderma* sp. isolates compared to the control.

* Averages followed by the same letter for each time evaluated do not differ according to Tukey's test ($p < 0.05$).

Similarly, the TO 245 isolate of *Fusarium* sp. was inhibited only up to 48 hours of growth under the gases released by the T3 isolate of *Trichoderma* sp. After 120 hours, this difference was no longer significant in relation to the control group (Figure 19).

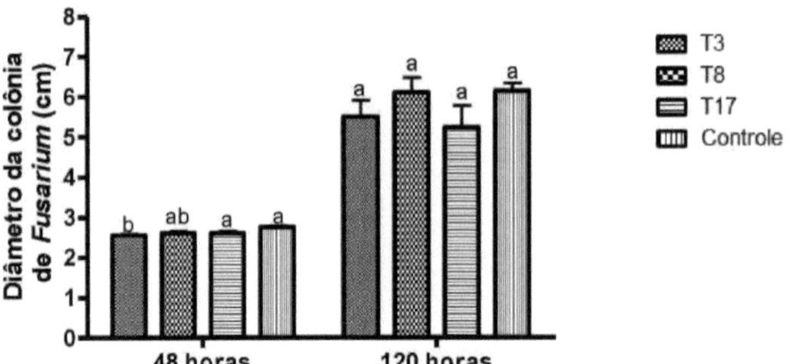

Figura 19.   Colony diameter of *Fusarium* sp. isolate TO 245 in an inhibition test by volatile metabolites released by *Trichoderma* sp. isolates in relation to the control.

* Averages followed by the same letter for each time evaluated do not differ according to Tukey's test ($p <$ 0.05).

Although in this study a variable inhibition of the development of colonies of strain 26380 belonging to Fol race 3 was observed in direct confrontation with isolates of *Trichoderma* sp. (Table 5), in the treatment with volatile metabolites there was no significant inhibition in relation to the control, as can be seen in Figure 20.

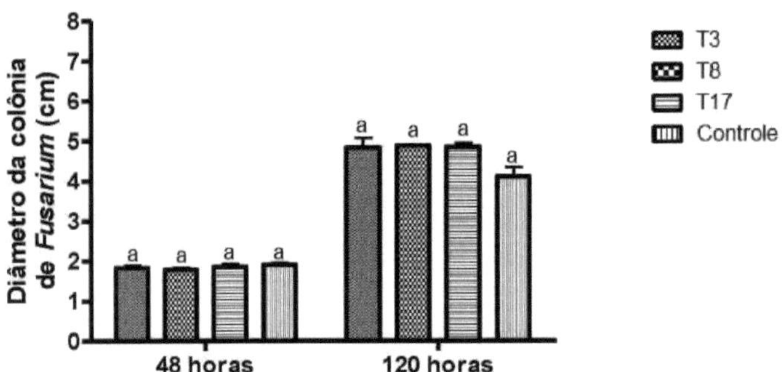

Figura 20.   Diameter of colonies of *Fusarium* sp. strain 26380 in inhibition test by volatile metabolites released by *Trichoderma* sp. isolates in relation to control

* Averages followed by the same letter for each time evaluated do not differ according to Tukey's test ($p <$ 0.05).

It was observed, however, that in the treatment with isolate T8 of *Trichoderma* sp. there was an inhibition of the sporulation of the colonies of the *Fusarium* sp. isolates, with sporulation being observed in the non-confronted colonies of the control (Figure 21).

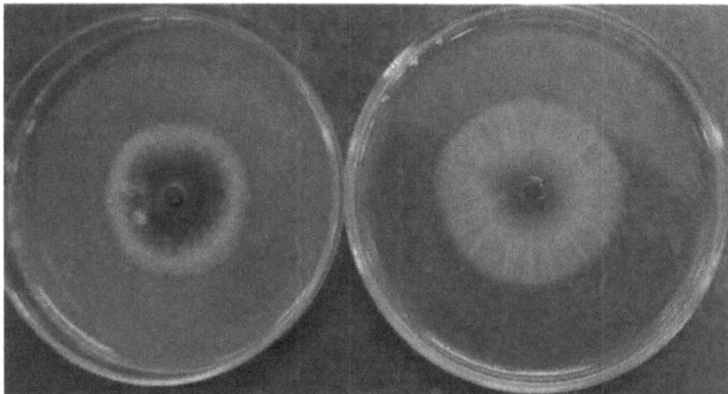

Figure 21. Comparison between colonies of strain 26380, control group (A), and in the presence of volatile metabolites released by *Trichoderma* sp. isolate T8 (B).

For strain 34970 there was significant growth inhibition when paired with each of the 3 isolates of *Trichoderma* sp. compared to the control after 120 hours of testing (Figure 22). The average diameter of growth paired with isolate T17 of *Trichoderma* sp. reached 3.5 cm compared to 4.9 cm for the control (Figure 23).

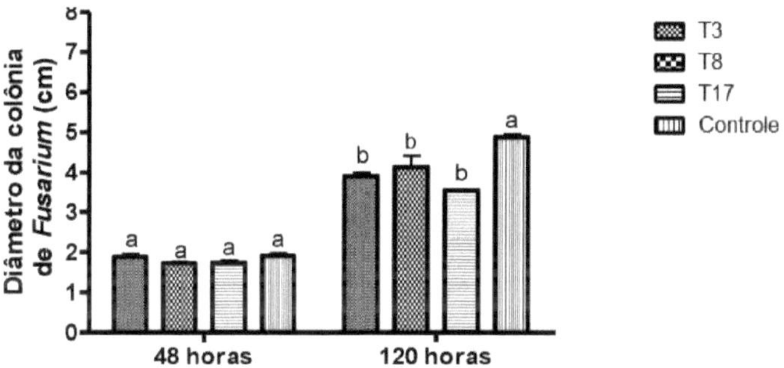

Figure 22. Diameter of colonies of *Fusarium* sp. strain 34970 in an inhibition test using volatile metabolites released by isolates of *Trichoderma* sp. in relation to the control.

* Averages followed by the same letter for each time evaluated do not differ according to Tukey's test ($p < 0.05$).

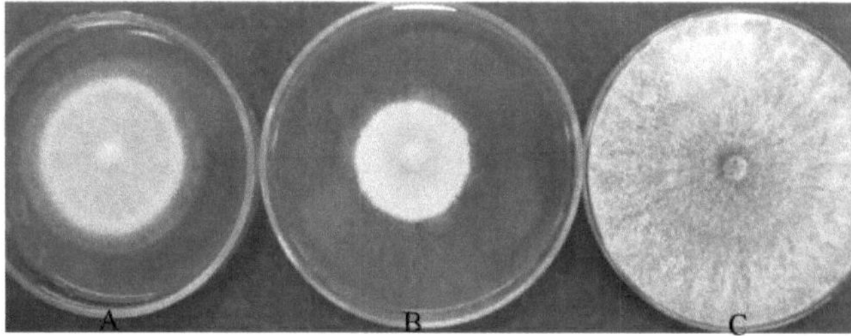

Figure 23. Colonies of strain 34970 tested for volatile metabolites released by *Trichoderma* sp. isolate T17 (A). In the center, tested strain (B) and control (C).

39

Variability can be expected in the antagonism capacity of species and even antagonist isolates within the *Trichoderma* genus, and likewise for the resistance capacity of *Fusarium* sp. isolates, *since* these characteristics are complex and determined by several genes or sets of genes, both in the antagonist and in the pathogen.

Figure 24 shows the data obtained in relation to the growth of *Fol* strain OSU 451 in the presence of the *Trichoderma* sp. isolates, which demonstrate the absence of any effect on the development of the target pathogen by any of the *Trichoderma* sp. isolates evaluated.

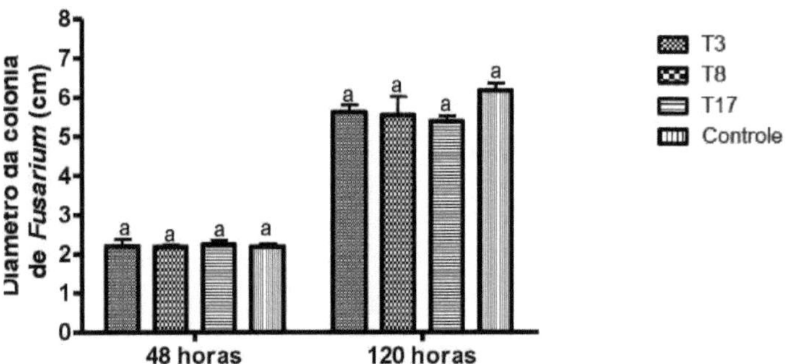

Figura 24. Colony diameter of the OSU 451 strain of *Fusarium* sp. in an inhibition test by volatile metabolites released by *Trichoderma* sp. isolates in relation to the control.

* Averages followed by the same letter for each time evaluated do not differ according to Tukey's test ($p < 0.05$).

Figure 25 shows that there was no statistical difference between the growth of the MM 66 Fol isolate and any of the *Trichoderma* sp. isolates in both evaluations.

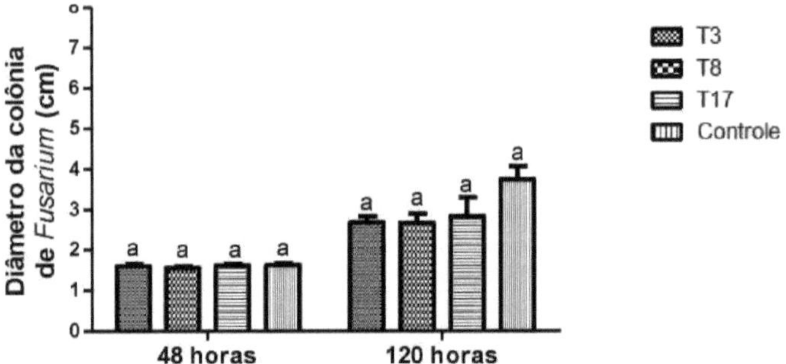

Figura 25. Colony diameter of *Fusarium* sp. strain MM 66 in an inhibition test using volatile metabolites released by *Trichoderma* sp. isolates compared to the control.

* Averages followed by the same letter for each time evaluated do not differ according to Tukey's test ($p < 0.05$).

The data presented in Figure 26 show that isolate 1205/2 of *Fusarium* sp. was not inhibited by any of the *Trichoderma* sp. isolates challenged during the entire experiment.

In a study by Küçük & Kivanç (2003), volatile and non-volatile metabolites from *T. harzianum* strains were tested against various phytopathogenic fungi, and it was observed that isolates of *Fusarium* spp. showed greater resistance than other fungi such as *Rhizoctonia* sp. and *Sclerotium* sp. The variability of the responses obtained in the evaluations of inhibition caused by competition and the production of volatile metabolites between the isolates in this study reinforces the results obtained by the authors, since three races of *Fusarium sp.* were evaluated against ten isolates of *Trichoderma* sp., which also showed variable behavior in relation to the isolates of *Fusarium* sp.

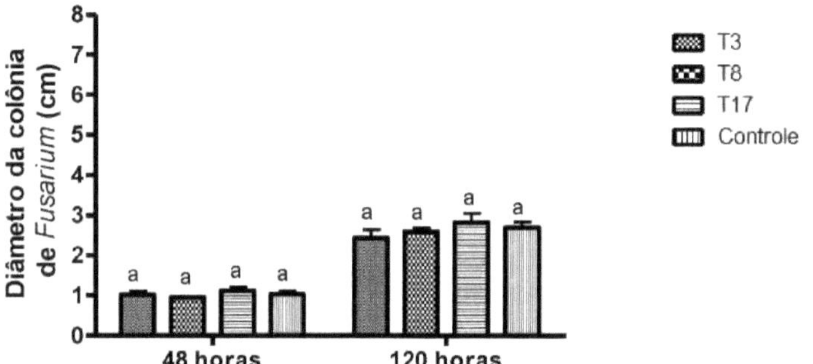

Figura 26.   Colony diameter of isolate 1205/2 of *Fusarium* sp. in an inhibition test by volatile metabolites released by isolates of *Trichoderma* sp. in relation to the control group.

* Averages followed by the same letter for each evaluation do not differ according to Tukey's test (p< 0.05).

The low inhibitory capacity of *Trichoderma* sp. isolates through the production of volatile metabolites does not necessarily indicate that they cannot be used to control diseases *in vivo*. Various other mechanisms such as competition for iron (Segarra *et al.* 2009) and nutrients, siderophore production, colonization of the plant's rhizosphere (Benitez *et al.* 2004) and induction of resistance (Woo *et al.* 2006) have already been described in the literature as key factors in combating plant disease-causing agents. In addition, authors such as Bélanger *et al.* (1995) report that the most promising strains of *Trichoderma* sp. to be applied in the biological control of plant diseases are those that produce non-volatile metabolites.

The ability of *Trichoderma* sp. to biocontrol plant diseases is not, however, restricted to the methods evaluated in this work. The coordinated action of several mechanisms at the same time makes it very difficult for phytopathogenic strains to emerge that are resistant to their control and can guarantee their efficiency in controlling diseases in the field. Thus, even if one control mechanism is predominant for an isolate, this does not exclude the possibility that one or more of the other mechanisms may play an important role in the antagonistic behavior of a particular isolate of *Trichoderma* sp.

4.4 Biological control

Forty-five days after planting the tomato seedlings in substrate treated with *Trichoderma* sp. and contaminated with isolates of *Fusarium* sp., it was possible to observe symptoms of disease in the plants from the treatments with substrate infested with isolates TO 245 and 1205/2 of *Fol*, as can be seen in Figure 27. The diseased plants showed symptoms of wilting to varying degrees on the severity scale, as well as yellowing or necrotic leaves with consequent wilting in some cases.

In the evaluation of the biological control of the TO 245 isolate of *Fusarium sp.*, the seedlings treated with the T6 and T17 isolates of *Trichoderma* sp. showed a lower degree of disease severity, with scores equal to 1.25 and 0.50 respectively, compared to 1.87 for the control treatment without *Trichoderma* sp., according to the scale of Vakalounakis *et al.* (2004) (Table 7).

Figure 27. Biological control tests A- Tomato plant showing signs of death from fusariosis in the treatment containing only isolate TO 245 from Fol. B- Healthy plant with isolate T17 of *Trichoderma* sp.in substrate containing isolate TO 245 of Fol.

In plants treated with isolate 1205/2 of *Fusarium* sp., the symptoms observed were less severe (Table 7). The treatments with isolates T6 and T17 achieved scores of 0.50 and 0.38 respectively, according to the severity scale, compared to 0.50 for the control treatment without the presence of *Trichoderma* sp.

Table 7: Disease severity index in the biological control test of isolates TO 245 and 1205/2 of *Fusarium* sp. by isolates T6 and T17 of *Trichoderma* sp. according to the scale of Vakalounakis *et al.* (2004): (0) - no symptoms; (1) - mild or moderate wilting, slight vascular discoloration of the stem; (2) - severe wilting and vascular discoloration; (3) - dead plant.

Fusarium isolates	*Trichoderma* isolates		
	T6	T17	Control
TO 245	1,25	0,5	1,87
1205/2	0,5	0,375	0,5

These results corroborate those of Luongo *et al.* (2005), who analysed the ability of *Fusarium* sp. to colonize wheat and maize in the presence of antagonists such as *Trichoderma* sp. with moderate and significant reductions in the disease.

Another study with similar results, in which the effectiveness of reducing fusarium wilt in tomato was evaluated (Larkin & Fravel 1998), used isolates of *T. hamatum which* reduced the incidence of the disease by up to 64%. In addition to these hosts, *Trichoderma* sp. has been reported parasitizing *sclerotia of Sclerotinia sclerotiorum* (Lib.) and reducing the incidence of this disease in tomato by up to 88% (Abdullah *et al.* 2008).

The control of fusarium by antagonistic organisms involves mechanisms that restrict the growth and extent of colonization of the pathogen, either physically by creating barriers, or by inducing the production of phytoalexins in the plant (De Cal *et al.* 2000).

In the plants that showed signs of disease, it was possible to observe the obstruction of the stem vessels, as confirmation of death by fusarium wilt, compared to the treated plants and those without disease symptoms with unobstructed vessels (Figure 28).

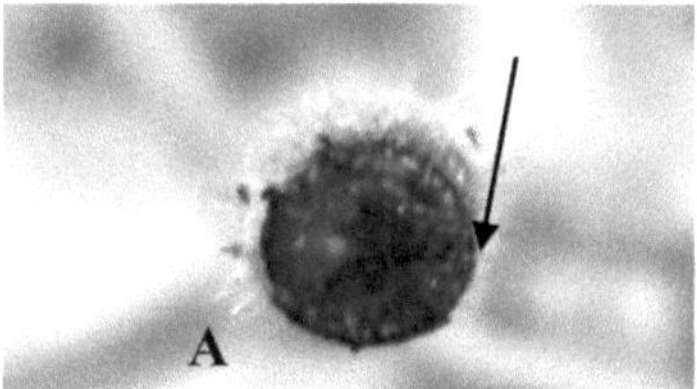
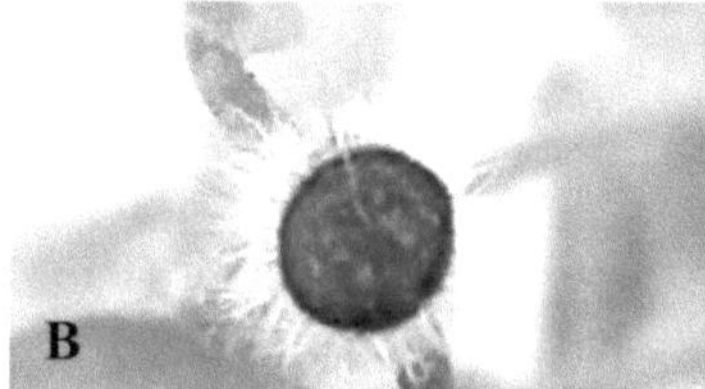

Figure 28. Cross-section of the stem of a tomato plant with clogged pots (A) highlighted, and a cross-section of a healthy tomato plant without clogged pots (B).

Biological control can be an efficient alternative to the use of cultivars resistant to races of Fol, since this method, in addition to exerting constant selection pressure for new races, is costly for small farmers since tomato seeds that are resistant to races of this pathogen are more expensive.

Biological treatment with *Trichoderma* sp. species can stimulate plant development, but under the conditions of this study this result was not observed. The height of the plants treated with *Trichoderma* sp. showed no significant difference for either treatment (Figure 29). Similarly, Ethur *et al.* (2008) found no differences between treatments with *T. harzianum* in the substrate and tomato seeds in relation to the development and vigor of the seedlings.

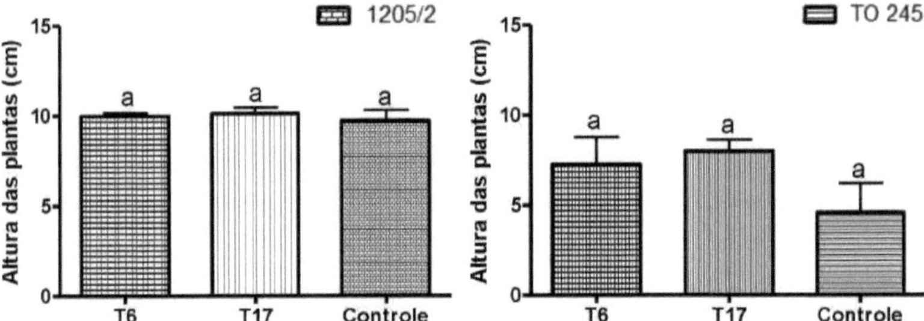

Figura 29. Height of tomato plants in the biological control test treated with *Trichoderma* sp. isolates T6 and T17, compared to the control treatment containing only Fol 1205/2 and TO 245 isolates.

The length of the roots of the tomato plants also showed no significant difference compared to the control treatment not treated with *Trichoderma* sp. (Figure 30).

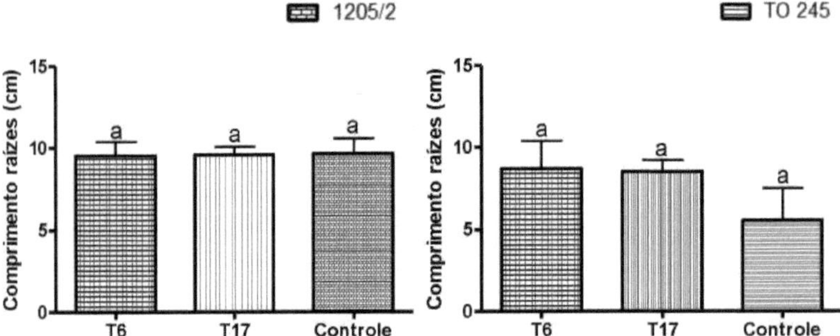

Figura 30. Root length of tomato plants in a biological control test treated with *Trichoderma* sp. isolates T6 and T17, compared to the control treatment containing only Fol 1205/2 and TO 245 isolates.

The dry weight of the aerial part of the tomato plants treated with isolates T6 and T17 of *Trichoderma* sp. was 0.044g and 0.039g respectively, compared to 0.031g for the control (Figure 31). For the substrate inoculated with isolate TO 245 and the *Trichoderma* sp. isolates, no significant differences were found in any of the treatments and in relation to the control treatment.

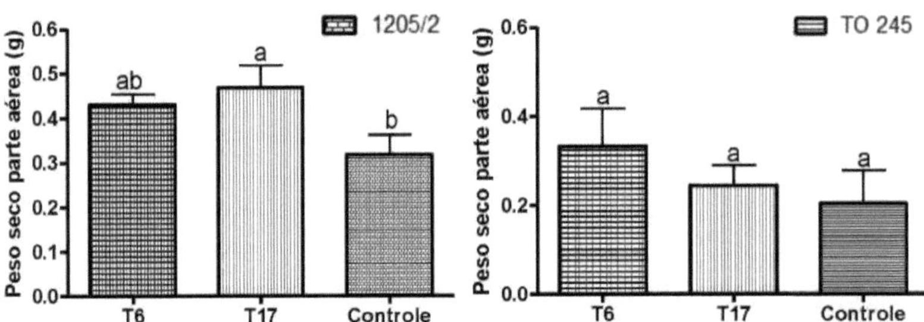

Figura 31. Dry weight of the aerial part of tomato plants in the biological control test treated with isolates

T6 and T17 of *Trichoderma* sp., compared to the control treatment containing only isolates Fol 1205/2 and TO 245.

The dry weight of the roots showed no statistical difference in any of the treatments carried out for any of the substrates inoculated with Fol isolates (Figure 32). Similarly, using antagonist inoculation methods to control tomato fusariosis, González *et al.* (2004) did not obtain results that differed from the control treatment in a test using only *Trichoderma* sp. The best results were obtained when combining this with a chemical soil treatment process.

Different results were found by Avendano *et al.* (2006), who showed a significant increase in root volume and in the number of bean rootlets when treated with *Trichoderma* sp. in conjunction with *Pseudomonas fluorescens*. The application of combined biological control agents is another possibility to be considered, since by acting with different biocontrol mechanisms, they can guarantee greater efficiency in the field.

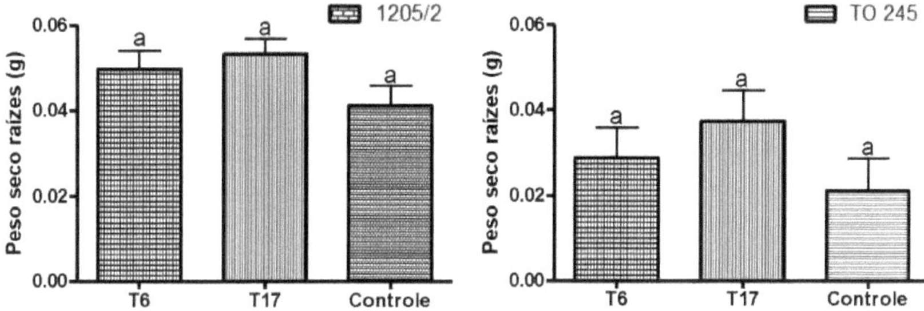

Figura 32. Dry weight of the roots of tomato plants in a biological control test treated with isolates T6 and T17 of *Trichoderma* sp., compared to the control treatment containing only isolates Fol 1205/2 and TO 245.

The application of *Trichoderma* sp. isolates in Integrated Management is one of the possibilities to be considered in the biocontrol of diseases. It is a viable alternative for reducing the use of fungicides and growth regulators, which in addition to reducing the environmental impact of this form of control can also reduce production costs.

4.5 Plant Compatibility Groups

The generation and selection of mutants capable of utilizing chlorate and not nitrate showed the generation of faster growth sectors for all Fol isolates.

Figure 33 shows the development of the Fusarium 23 strain with the sectors in BDC medium.

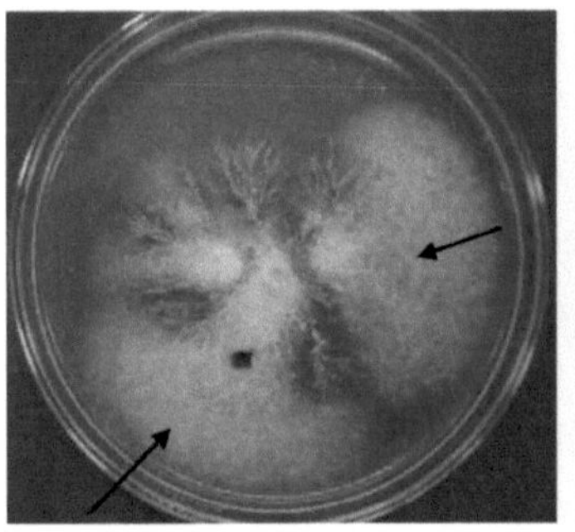 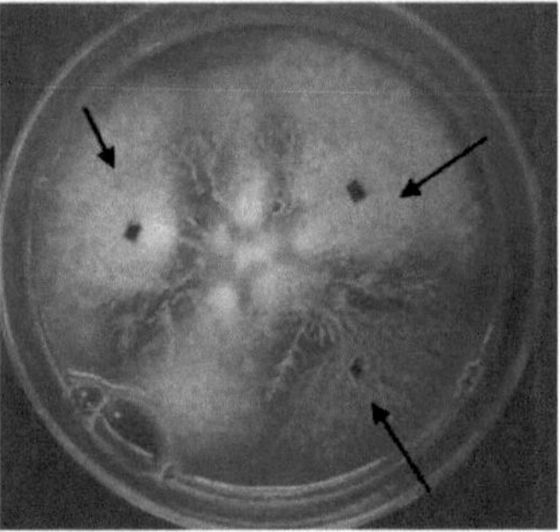

Figure 33. Colonies of the isolate Fusarium 23 grown on BDC medium. Highlighted are sectors of rapid growth.

Among the sectors replanted on minimal medium containing nitrate as the only nitrogen source, many colonies showed growth similar to the parental, indicating a possible reversal in one of the genes responsible for the mutant phenotypes. Figure 34 shows the mutants with extremely sparse mycelium compared to the regular-growing parent with dense mycelium.

A 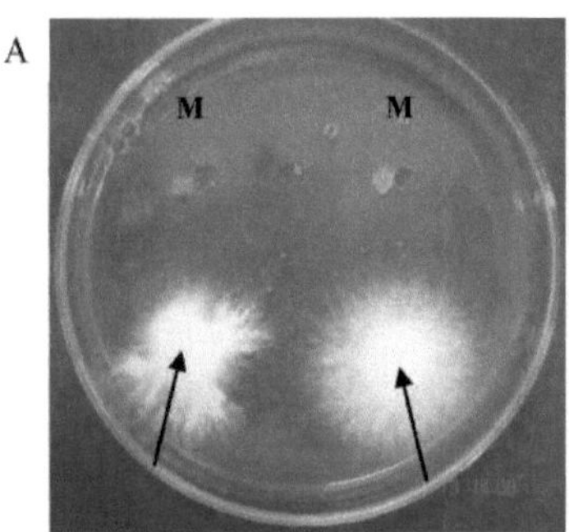 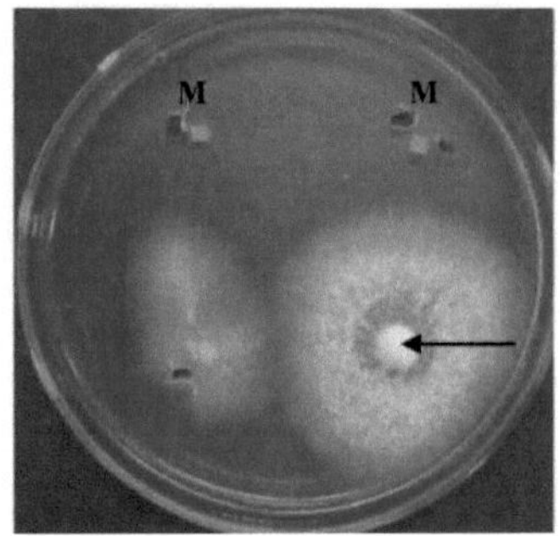 B

Figure 34. Colonies of isolates TO 11 (A) and 34970 (B) grown on MM medium. Highlighted are sectors of growth similar to the parent. Marked with the letter M, mutant colonies.

After this, the colonies that grew sparse mycelium from the same parent were considered mutants and were replanted on MM medium to check for heterokaryosis between them (Figure 35).

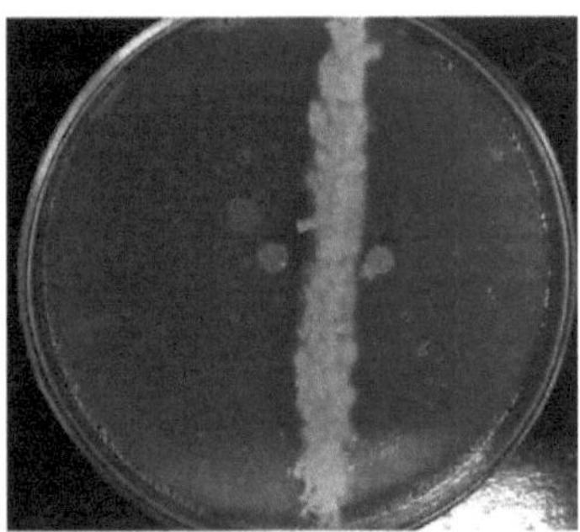

Figure 35. Heterozygous mutants of strain 26380.

After generating and selecting nit1 and NitM mutants in a medium containing nitrite and hypoxanthine, they were paired in all possible combinations with mutants from the standard strains of each of Fol's 4 GCVs.

The results of heterokaryon formation with the standard strains were positive in relation to the mutants belonging to the standard strain 34970 corresponding to GCV 0030, for the mutants referring to Fusarium isolates 23 and 27, TO 11 and 245 and 1205/2 (Table 8).

Table 8. Race and vegetative compatibility group of *Fol*.

Isolated	Race	VCG
1205/2	2	0030
Fusarium 23°	2	0030
Fusarium 27°	1	0030
TO 11°	2	0030
TO 245°°	2	0030

° Provided by Dr. Ailton Reis of Embrapa Hortaliças. °° Provided by Dr. Rômulo Kobori of Sakata Seeds through Dr. Wagner Bettiol of Embrapa Meio Ambiente.

One of the possible links to Fol is that isolates belonging to GCV 0030 have been associated with tomato plants for a fairly long period of time. One of the evidences of the relationship between this group and its host is due to the fact that all races are present in it, an indication of the adaptation of the fungal population to changes in the host's resistance spectrum (Gale *et al.* 2003).

Obtaining isolates from different locations in the same GCV indicates that this genetically homogeneous population is dispersed among geographically different regions. This type of result may indicate a genetic similarity between isolates and a probable origin from a common ancestor. Some studies indicate that isolates belonging to the same GCV commonly have identical multilocus haplotypes and belong to the same clonal lineage (Kistler *et al.* 1998). The occurrence of a GCV in more than one region can be explained by a wide initial distribution from a common geographical origin through seeds, seedlings, contaminated apparatus and human activities (Ahn *et al.* 1998).

A commonly accepted hypothesis that explains the presence of isolates belonging to different races in the same GCV in *F. oxysporum* f. sp. *lycopersici*, as observed for Fusarium isolates 23 and 27 of races 2 and 1 respectively, is based on the derivation of race 2 by race 1, and race 3 by race 2, explaining the presence of the 3 races in GCV 0030 for example (Kawabe *et al.* 2005). The diversity of races within the same GCV, as observed in the present work, implies that new races can arise independently in different localities as well as in a center of origin, followed by dispersal over long distances. This hypothesis, however, remains speculative, as there is still little knowledge about the stability of *vic* and virulence loci in pathogenic forms of *F. oxysporum* (Elias & Schneider, 1991).

The limited number of isolates used in this work prevents a definitive conclusion regarding the importance of heterokaryon formation in influencing genetic diversity and gene flow in *F. oxysporum* f. sp. *lycopersici in* Brazil. There is therefore a need for further studies including isolates of race 3 and from different parts of the country in order to broaden the scope of the study.

5. CONCLUSIONS AND PROSPECTS

- Isolates 589, 859, 921, 921/2 and 1205/1 of *Fusarium* sp. did not prove to be pathogenic to any of the differentiating tomato cultivars used.

- None of the tested isolates of *Fusarium* sp. showed pathogenicity on cultivars resistant to races 1 and 2, and race 3.

- In the direct confrontation tests, the lowest values obtained in the evaluation according to the Bell *et al.* scale were obtained with *Trichoderma* sp. isolates T3, T8 and T17, with colony overlap of at least six of the nine *Fol* isolates confronted.

- *Fol* strain 34970 was little inhibited by the *Trichoderma* spp. isolates, and isolate TO 11 had the most inhibited growth. Isolates T1 and T3 of *Trichoderma* sp. showed the highest overall averages in terms of inhibition capacity.

- Three of the *Fol isolates* were inhibited by the volatile metabolites produced by *Trichoderma* sp. isolates T3, T8 and T17.

- The results of biological control with *Trichoderma* sp. isolates T6 and T17 on tomato plants grown in substrate infested with *Fol* isolates TO 245 and 1205/2 showed lower disease severity in both cases.

- A statistical difference was observed in the dry weight of the aerial part when treated with the suspension of *Trichoderma* sp. isolate T17, with no statistical difference observed in the other treatments and tests.

- All the *Fol* isolates were grouped together in VCG 0030, indicating a possible genetic similarity between them.

PERSPECTIVES

The following are some perspectives for further research into the data obtained in this study:

1. Sequencing the ITS rDNA region of the *Fol* isolates in order to establish similarity relationships between the isolates and complement the results obtained in the vegetative compatibility tests.

2. To identify *Trichoderma* sp. isolates to species level using molecular biology techniques.

3. Carry out field tests to observe the protective effects of isolate T17 of *Trichoderma* sp. on tomato crops.

4. To establish the possible origin or emergence of race 3 in the country, by analyzing the vegetative compatibility groups of isolates belonging to it.

5.   To establish the vegetative compatibility groups of other *special forms* of *F. oxysporum* belonging to the collection of the Laboratory of Biological Control of Plant Diseases at the University of Caxias do Sul.

6. BIBLIOGRAPHICAL REFERENCES

Abdell-Fattah, G.M.; Shabana, Y.M.; Ismail, A.E.; Rashad, Y.M. (2007) *Trichoderma harzianum*: a biocontrol agent against *Bipolaris oryzae*. **Mycopathologia** 164: 81-89.

Abdullah, M.T.; Ali, .Y.; Suleman, P. (2008) Biological control of *Sclerotinia sclerotiorum* (Lib.) de bary with *Trichoderma harzianum* and *Bacillus amyloliquefaciens*. **Crop Prot** 27: 1354-1359.

Ahn, I. P.; Chung, H. S.; Lee, Y-H. (1998). Vegetative compatibility groups and pathogenicity among isolates of *Fusarium oxysporum* f. sp. *cucumerinum*. **Plant Dis** 82: 244-246.

Alfenas, A. C.; Mafia, R. G. (2007) Methods in Phytopathology. Editora UFV 382p

Andrade, E. R.; Schuck, E.; Dal Bó, M. A. (1993) Evaluation of the resistance of *Vitis* spp. to *Fusarium oxysporum f. sp. herbemontis* under controlled conditions. **Pesq Agropec Bras** 28: 1287-1290.

Armstrong, G. M. & Armstrong, J. K. (1981). *Formae speciales* and races of *Fusarium oxysporum* causing wilt diseases. In: Nelson, PE. Tousson, T.A. & Cook, R.J. (Eds) **Fusarium: Disease, Biology and Taxonomy**. Pennsylvania State University Press, University Park. 1981. pp 391-399.

Avendano, C.; Arbelàez, G.; Rondón, G. (2006) Control biològico del marchitamineto vascular causado por *Fusarium oxysporum* f sp. *phaseoli* em frijol *Phaseolus vulgaris* L., mediante la acción combinada de *Entrosphora colombiana*, *Trichoderma* sp. y *Pseudomonas fluorescens*. **Agron Colomb** 24: 62-67.

Belabid, L.; Fortas, Z. (2002) Virulence and vegetative compatibility of Algerian isolates of *Fusarium oxysporum* f. sp. *lentis*. **Phytopathol Mediterr** 41: 179-187

Bélanger, R. R.; Dufour, N.; Caron, J.; Benhamou, N. (1995) Chronological events associated with the antagonistic properties of *Trichoderma harzianum* against *Botrytis cinerea*: indirect evidence for sequential role of antibiosis and parasitism. **Bioc Sci Tech** 5: 41-53.

Bell, D. K.; Wells, H. D.; Markham, C. R. (1982) *In vitro* antagonism of *Trichoderma* species against six fungal pathogens. **Phytopathology** 72: 370-382.

Benitez, T.; Rincón, A. M.; Limón, M. C.; Codón, A. C. (2004) Biocontrol mechanisms of *Trichoderma* strains. **Int Microbiol** 7: 249-260.

Carvalho, A. O.; Neto, J. J.; do Carmo, M. G. F. (2005) Colonization of tomato roots by *Fusarium oxysporum* f sp. *lycopersici* in nutrient solution with three nitrogen sources. **Fito Bras** 30: 26-32.

Cai, G.; Gale, L. R.; Schneider, R. W.; Kistler, H. C.; Davis, R. M.; Elias, K. S.; Miyao, E. M. (2003) Origin of race 3 of *Fusarium oxysporum* f. Sp. *lycopersici* at a single site in California.

Phytopathology. 93: 1014-1022.

Correl, J. C.; Klittich, C. J. R.; Leslie, J. F. (1987) Nitrate non-utilizing mutants of *Fusarium oxysporum* and their use in vegetative compatibility tests. **Phytopathology** 77: 1640-1646.

Cove, D. J. (1976) Chlorate toxicity in *Aspergillus nidulans*; Studies of mutants altered in nitrate assimilation. **Mol Gen Genet** 146: 147-159.

Dal Bello, G. M.; Mònaco, C. I.; Châves, A. R. (1997) Efecto de los metabolitos volâtiles de *Trichoderma hamatum* sobre el crecimiento de hongos fitopatógenos procedentes del suelo. **Rev Iberoam de Micol** 14: 131-134.

De Cal, A.; Garcia-Lepe, R.; Melgarejo, P. (2000). Induced resistance by *Penicillium oxalicum* against *Fusarium oxysporum* f. sp. *lycopersici*: histological studies of infected and induced tomato stems. **Phytopathology** 90: 260-268.

Dennis, C.; Webster, J. (1971a). Antagonistic properties of species-groups of *Trichoderma* I. Production of non-volatile antibiotics. **Trans Br Mycol Soc** 57: 25-39.

Dennis, C.; Webster, J. (1971b). Antagonistic properties of species-groups of *Trichoderma*. II. Production of volatile antibiotics. **Trans Br Mycol Soc** 57: 41-48.

De Oliveira, V. C. & Da Costa, J. L. S. (2003) Vegetative compatibility of nit-mutants of *Fusarium solani* pathogenic and non-pathogenic to bean and soybean. **Fito Bras** 28: 89-92.

Elias, K.S.; Schneider, R.W. (1991) Vegetative compatibility groups in *Fusarium oxysporum* f. sp. *lycopersici*. **Phytopathology** 81: 159-162.

Embrapa, Hortaliças (2008). Production Systems. **Available (online)** http://sistemasdeproducao.cnptia.embrapa.br/FontesHTML/Tomate/TomateIndustrial_2ed/doe ncas_fungo.htm (October 31).

Ethur, L.Z.; Blume, E.; Muniz, F.B.; Camargo, R.F.; Flores, M.G.V.; Cruz, J.L.G.; Menezes, J.P. (2008) *Trichoderma harzianum* in the development and protection of seedlings against tomato fusariosis. **Ciência e Natura** UFSM 30: 57-69.

Fernandez, D.; Assigbetse, K.; Dubois, M.-P.; Geiger, J.-P. (1994) Molecular characterization of races and vegetative compatibility groups in *Fusarium oxysporum* f. sp. *vasinfectum*. **Appl Environ Microbiol** 60: 4039-4046.

Fischer, I. H.; Almeida, A. M.; Garcia, M. J. D. M. (2006) Effect of fungicides on the mycelial growth of *Fusarium subglutinans in vitro*. In: 19th Annual Meeting of the Biological Institute. **Expanded Abstract.** Pp. 68: 1-4.

Flor, H. H. (1971) Current status of the gene-for-gene concept. **Annu Rev Phytopathol.** 9: 275296.

Forsyth, L. M.; Smith, L. J.; Aitken, E. A. B. (2006) Identification and characterization of non-pathogenic *Fusarium oxysporum* capable of increasing and decreasing *Fusarium* wilt severity. **Mycol Res** 110:929-935.

Gale, L. R.; Katan, T.; Kistler, H. C. (2003) The probable center of origin of *Fusarium oxysporum* f. sp. *lycopersici* VCG 0030. **Plant Dis** 87: 1433-1438.

Galli, F. (1980) Manual de Fitopatologia. Diseases of Cultivated Plants. Agronômica Ceres. Vol. 2. 587p.

Ghisalberti, E. L. & Sivasithamparam, K. (1991) Antifungal antibiotics produced by *Trichoderma* spp. **Soil Biol Bioch** 23: 1011-1022.

Gómez, I.; Chet, I.; Herrera-Estrella, A. (1997) Genetic diversity and vegetative compatibility groups among *Trichoderma* isolates. **Mol Gen Genet** 256: 127-135.

Gónzalez, R.; Montealegre, J.; Herrera, R. (2004) Control Biológico de *Fusarium solani* en tomate mediante empleo de los bioantagonistas *Paenibacillus lentimorbus* y Trichoderma spp. **Cien e Inv Agr** 31:21-28.

Graph Pad Software, Inc. USA. February 2009. CD-ROM.

Grattidge, R.; O' Brien, R. G. (1982) Occurrence of a third race of *Fusarium* wilt of tomatoes in Queensland. **Plant Dis** 66: 165-166.

Harman, G. E. (2000) Myths and Dogmas of Biocontrol. Changes in the perceptions derived from the research on *Trichoderma harzianum* T-22. **Plant Dis** 84:377-393.

Harman, G. E.; Howell, C. R.; Viterbo, A.; Chet, I.; Lorito, M. (2004) *Trichoderma* species-opportunistic, avirulent plant symbionts. **Nature** 2: 43-56.

Harman, G. E. (2006) Overview of mechanisms and uses of *Trichoderma* spp. **Phythopathology** 96:190- 194.

Hibar, K.; Daami-Remadi, M.; Khiareddine, H.; El Mahjoub, M. (2005) Effet inhibiteur in vitro et in vivo du *Trichoderma harzianum* sur *Fusarium oxysporum* f. sp. *radicis-lycopersici*. **Biotechnol Agron Soc Environ.** 9: 163-171.

Howell, C. R. (2003) Mechanisms employed by *Trichoderma* species in the biological control of plant diseases: the history and evolution of current concepts. **Plant Dis** 87:4-10.

Howell, C. R. (2006) Understanding the mechanisms enployed by *Trichoderma virens* to effect biological control of cotton diseases. **Phytopathology** 96: 178- 180.

Jacobson, D. J. & Gordon, T. R. (1988) Vegetative compatibility and self-incompatibility within *Fusarium oxysporum* f. sp. *melonis*. **Phytopathology** 78: 668-672.

Katan, T.; Zamir, D.; Sarfatti, M.; Katan, J. (1991) Vegetative compatibility groups and subgroups in *Fusarium oxysporum* f. sp. *radicis-lycopersici*. **Phytopathology** 81:255-262.

Katan, T & Di Primo, P. (1999) Current status of vegetative compatibility groups in *Fusarium oxysporum*: Supplement. **Phytoparasitica** 27:1-7.

Kawabe, M.; Kobayashi, Y.; Okada, G.; Yamaguchi, I.; Teraoka, T.; Arie. (2005) Three evolutionary lineages of tomato wilt pathogen, *Fusarium oxysporum* f. sp. *lycopersici,* based on sequences of IGS, MAT1, and pg1, are each composed of isolates of a single or closely related vegetative compatibility group. **J Gen Plant Pathol** 71: 263-272.

Kimati, H.; Amorim, L.; Rezende, J.A.M.; Bergamim Filho, A.; Camargo, L.E.A. (2005) Manual de Fitopatologia Volume 2 Doenças das plantas cultivadas. Editora Agronômica Ceres 663p.

Kistler, H. C. (1997) Genetic diversity in the plant-pathogenic fungus *Fusarium oxysporum*. **Phytopathology** 87: 474-479.

Kistler, H. C.; Alaubovette, C.; Baayen, R. P.; Bentley, S.; Brayford, D.; Coddington, A.; Correll, J.; Daboussi, M.-J.; Elias, K.; Gordon, T. R.; Katan, T.; Kim, H. G.; Leslie, J. F.; Martyn, R. D.; Migheli, Q.; Moore, N. Y.; O' Donnell, K.; Ploetz, R. C.; Rutherford, M. A.; Summerell, B.; Waalwijk, C.; Woo, S. (1998) Systematic numbering of vegetative compatibility groups in the plant pathogenic fungus *Fusarium oxysporum*. **Phytopathology** 88:30- 32.

Klittich, C. J. R.; Leslie, J. F. (1988) Nitrate reduction mutants of *Fusarium moniliforme* (*Giberella fujikuroi*). **Genetics** 118: 417-423.

Kopacki, M. & Wagner, A. (2006) Effect of some fungicides on mycelium growth of *Fusarium avenaceum* (Fr.) Sacc. pathogenic to chrysanthemum (*Dedranthemma grandiflora* Tzvelev) **Agron Res** 4: 237-240.

Küçük, C. & Kivanç, M. (2003) Isolation of *Trichoderma* spp. and determination of their antifungal, biochemical and physiological features. **Turk J of Biol** 27: 247- 253.

Küçük, C. & Kivanç, M. (2004) In vitro antifungal activity of strains of *Trichoderma harzianum*. **Turk J of Biol** 28-111-115.

Larkin, R. P. & Fravel, D. R. (1998) Efficacy of various fungal and bacterial biocontrol organisms for control of *Fusarium* wilt of tomato. **Plant Dis** 82: 1022-1028.

Leslie, J. F. (1996) Fungal Vegetative Compatibility- Promises and Prospects. **Phytoparasitica** 24: 3-6.

Leslie, J. F. & Zeller, K.A. (1996) Heterocaryon compatibility in fungi - more than just another way to die. **J Genet** 3: 415-424.

Louzada, G. A.; Lobo Junior, M.; Marchao, R. L.; Balbino L. C. (2004) Effect of crop rotation on *Fusarium* spp. and microbiological activity in a crop-livestock integration area. **Available (online)** http://sistemasdeproducao.cnptia.embrapa.br/ embrapaintegracaolavourapecuaria.

Luongo, L.; Galli, M.; Corazza, L.; Meekes, E.; de Haas, L.; Plas, C.L.V.D.; Kohl, J. (2005) Potential of fungal antagonists for biocontrol of *Fusarium* spp. in wheat and maize trough competition in crop debris. **Biocontrol Sci Techn** 3: 229-242.

McCallum, B. D.; Tekauz, A.; Gilbert, J. (2001) Vegetative compatibility among *Fusarium graminearum (Gibberela zeae)* islates from barley spikes in southern Manitoba. **Can J of Plant Pathol** 23: 83-87.

Melo, L. S.; & Valarini, P. J. (1995) Potential of rhizobacteria in the control of *Fusarium solani* (Mart.) Sacc. on *cucumber* (*Cucumis sativum* L.). **Sci Agricola** 52: 326-330.

Mes, J. J.; Van Doorn, J.; Roebroeck, E. J. A.; Van Egmond, E.; Van Aartrijk, J.; Boonekamp, P. M. (1994) Restriction fragment length polymorphisms, races and vegetative compatibility groups within a worldwide collection of *Fusarium* f. sp. *gladioli*. **Plant Path** 43: 362-370.

Mes, J. J.; Westeijn, E. A.; Herlaar, F.; Lambalk, J. J. M.; Wijbrandi, J.; Haring, M. A.; Cornelissen, B. J. C. (1999). Biological and molecular characterization of *Fusarium oxysporum* f. sp. *lycopersici* divides race 1 isolates into separate virulence groups. **Phytopathology** 89: 156-160.

Monte, E. (2001) Understanding *Trichoderma*: between biotechnology and microbial ecology. **Int Microbiol** 4: 1-4.

Nechet, K. L.; Halfeld-Vieira, B. A.; Pereira, P. R. V. S. (2004) Diagnosis of banana diseases in the state of Roraima. **Research and Development Bulletin: Embrapa** Roraima November.

Ogiso, H.; Fujinaga, M.; Saito, H.; Takehara, T.; Yamanaka, S. (2002) Physiological races and vegetative compatibility groups of *Fusarium oxysporum* f. sp. *lactucae* isolated from crisphead lettuce in Japan. **J Gen Plant Pathol** 68: 292-299.

Paradela, A. L. (2002) In vitro behavior of *Fusarium* sp., *Rhizoctonia solani* and *Sclerotinia sclerotiorum,* against isolates of *Trichoderma* sp. *Technical report on Trichoderma.* **Natural Rural**: 1-5.

Puhalla, J. E. (1985) Clasification of strains of *Fusarium oxysporum* on the basis of vegetative compatibility. **Can J Botany** 62: 540-545.

Reis, A.; Boiteux, L. S.; Giordano, L. B.; Costa, H.; Lopes, C. A. (2004) Occurrence of *Fusarium oxysporum* f.sp. *lycopersici* race 3 in tomatoes in Brazil and selection of new sources of resistance to the pathogen. **Research and Development Bulletin** - Embrapa Hortaliças.

Reis, A.; Costa, H.; Boiteux, L. S.; Lopes, C. A. (2005) First report of *Fusarium oxysporum* f. sp. *lycopersici* race 3 on tomato in Brazil. **Fito Bras** 30: 426-428.

Reis, A.; Boiteux, L. S.; Urben, A. F.; Costa, H. (2006) Establishment and characterization at race level of a collection of isolates of *Fusarium oxysporum* f. sp. *lycopersici*. **Research and Development Bulletin 19** - Embrapa Hortaliças.

Reis, A & Boiteux, L. S. (2007). Outbreak of *Fusarium oxysporum* f. sp. *lycopersici* race 3 in commercial fresh-market tomato fields in Rio de Janeiro state, Brazil. **Hortic Bras** 25: 451454.

Sànchez, V.; Rebolledo, O.; Picaso, R.M.; Càrdenas, E.; Córdova, J.; Gonzâlez, O.; Samuels, G.J. (2007) In vitro antagonism of *Thielaviopsis paradoxa* by *Trichoderma longibrachiatum*. **Mycopathologia** 163: 49-58.

Santos, M. M. dos & Noronha, J. F. (2001) Diagnosis of table tomato cultivation in the municipality of Goianâpolis, state of Goiàs, Brazil. **Pesq Agropec Trop** 31: 29-34.

Santos, B. A.; Zambolim, L.; Ventura J. A.; Vale, F. X. R. (2002). Severity of isolates of *Fusarium subglutinans* f. sp. *ananas* sensitive and resistant to benomyl in pineapple. **Fito Bras** 27: 101- 103.

Segarra, G.; Casanova, E.; Avilés, M.; Trillas, I. (2009) *Trichoderma asperellum* strain T34 controls *Fusarium* wilt disease in tomato plant in soiless culture trough competition fpr iron. **Microb Ecol** available online http://www.springerlink.com/content/d5581727x37h2202

Shaw, I. M. & Taylor, A. (1986) The chemistry of peptides related to metabolites of *Trichoderma* spp. 2. An improved method of characterization of peptides of 2-methylalanine. **Can J Chem** 64: 164-173.

Silva, J. C. & Bettiol, W. (2005) Potential of non pathogenic *Fusarium oxysporum* isolates for control of *Fusarium* wilt of tomato. **Fito Bras** 30: 409-412.

Simao, R. & Rodriguez, T. D. M. (2008) Evolution of table tomato production in the state of Rondônia. **Anais XLVI Congresso da Sociedade Brasileira de Economia, Administraçâo e Sociologia Rural**.

Song, W.; Zhou, L.; Yang, C.; Cao, X.; Zhang, L.; Liu, X. (2004) Tomato *Fusarium* wilt and its chemical control strategies in a hydroponic system. **Crop Prot** 23: 243-247.

Stadnik, M. J. & Talamini, V. (2004) Manejo ecológico de doenças de plantas. Ed. Florianópolis, SC: CCA/ UFSC. 293p.

Swift, C. E.; Wickliffe, E. R.; Schwartz, H. F.; (2002) Vegetative compatibility groups of *Fusarium oxysporum* f. sp. *cepae* from onion in Colorado. **Plant Dis** 86: 606-610.

Vakalounakis, D. J & Fragdiaskis, G. A. (1999) Genetic diversity of *Fusarium oxysporum* isolates from cucumber: differentiation by pathogenicity, vegetative compatibility and RAPD fingerprinting. **Phytopathology** 89: 161-168.

Vakalounakis, D. J.; Wang, Z.; Fragkiadakis, G. A.; Skaracis, G. N.; Li, D.-B.   (2004) Characterization of *Fusarium oxysporum* isolates obtained from cucumber in China by pathogenicity, VCG and RAPD. **Plant Dis** 88: 645-649.

Valenzuela-Ureta, J. G.; Lawn, D. A.; Heisey, R. F.; Zamudio-Guzman, V. 1996. First report of *Fusarium* wilt race 3, caused by *Fusarium oxysporum* f.sp. *lycopersici,* of tomato in Mexico. **Plant Dis** 80: 105.

Vinale, F.; Ghisalberti, E. L.; Sivasithamparam, K.; Marra, R.; Scala, F.; Lorito, M. Secondary metabolites produced by *Trichoderma* spp. and their role interaction of this fungus with plants and other microorganisms. In: XLIX Italian Society of agricultural genetics annual congress. **Abstracts**. Pp. 12-15. Potenza, Italy. 2005.

Vinale, F.; Sivasithamparam, K.; Ghisalberti, E. L.; Marra, R.; Woo, S. L.; Lorito, M. (2008) *Trichoderma* -plant- pathogen interactions. **Soil Biol Biochem** 40: 1-10.

Woo, S.L.; Scala, F.; Ruocco, M.; Lorito, M. (2006) The molecular biology of the interactions between *Trichoderma* spp., phytopathogenic fungi, and plants. **Phytopathology** 96: 181- 185.

Zhihao, X.; Kaimay, H.; Weiling, S.; Sengjiu, Z.; Guojing, L.; Liping, C.; Zhihui, X.; Hongxia, W.; Xiuchu, X.; Guangliang, X.; Baishen, J. (2000) The determination of physiological race of *Fusarium oxysporum* f. sp. *lycopersici* of tomato in Zhejiang, China. **Acta Phys Plant** 22: 356- 358.

I want morebooks!

Buy your books fast and straightforward online - at one of world's fastest growing online book stores! Environmentally sound due to Print-on-Demand technologies.

Buy your books online at
www.morebooks.shop

Kaufen Sie Ihre Bücher schnell und unkompliziert online – auf einer der am schnellsten wachsenden Buchhandelsplattformen weltweit! Dank Print-On-Demand umwelt- und ressourcenschonend produziert.

Bücher schneller online kaufen
www.morebooks.shop

info@omniscriptum.com
www.omniscriptum.com

Printed by Books on Demand GmbH, Norderstedt / Germany